Magic Bullets, Miracle Drugs, and Microbiologists

T0131205

Magic Bullets, Miracle Drugs, and Microbiologists

A History of the Microbiome and Metagenomics

WILLIAM C. SUMMERS

ASM
PRESS
Washington, DC

WILEY

Copublication by the American Society for Microbiology and John Wiley & Sons, Inc.

Editorial Correspondence:
ASM Press, 1752 N Street, NW, Washington, DC 20036-2904, USA

Registered Offices:
John Wiley & Sons, Inc., 111 River Street, Hoboken, NJ 07030, USA

For details of our global editorial offices, customer services, and more information about Wiley products, visit us at www.wiley.com.

Wiley also publishes its books in a variety of electronic formats and by print-on-demand. Some content that appears in standard print versions of this book may not be available in other formats.

Library of Congress Cataloging-in-Publication Data applied for:

ISBN 9781683674771 (Paperback)

Cover Design: Wiley
Cover Image: © Cavan Images/Getty Images, luchschenF/Shutterstock

Set in 10.5/13pt ArnoPro by Straive, Pondicherry, India

SKY10075771_052424

Contents

List of Illustrations

Foreword

I've always thought of Bill Summers as a virologist. I was aware of his molecular biology research first on T7 bacteriophage and later, on herpes virus in collaboration with Wilma Summers. But on second thought, I realized that wasn't quite right. Summers is a historian of science as well, and as a historian his research and books include both viruses and bacteria. Examples are: Summers WC. 1999. *Félix d'Herelle and the Origins of Molecular Biology*; Summers WC. 2022. *The American Phage Group: Founders of Molecular Biology*; and Summers WC. 2012. *The Great Manchurian Plague, 1910–1911: Geopolitics of an Epidemic Disease* (all published by Yale University Press, New Haven, CT).

In this book he takes on the story of how the diversity of the microbial world has come to be understood from its discovery over four centuries ago to our current concept of microbiomes. Central to this understanding are the experimental methods used to identify, describe, and classify microbes. Summers argues that this understanding was revolutionized in the decades of the 1960s and 1970s when microbiologists adopted genetic approaches in place of the traditional physiological, morphological, and pathological approaches to identify and classify microbes.

The present book includes both bacteria and viruses (and at times a few other denizens of the microbial world). Bacteria often predominate but bacteriophage and animal viruses also have an important role throughout. Summers gives credit for the beginning of the subject of microbiology to van Leeuwenhoek's report of his microscope and what he saw in his letter to the Royal Society of 9 October 1675, but we are also reminded that Aristotle (384 BCE–322 BCE) should be acknowledged and given credit for his efforts to devise a comprehensive and rational classification scheme.

Once microbes were identified, Summers writes that "medical scientists sought microbial causes for every known malady, from cancer to mental disorders to heart disease and strokes. These hunters of microbes were remarkably successful. Many serious diseases of humankind turned out to be caused by microbes: tuberculosis, pneumonia, typhoid, cholera, polio, influenza... the list goes on and on." Now at the end of the 20th century and in the 21st century, research again implicates microbes in the initiation or cause of chronic diseases. *Helicobacter pylori* and human papillomavirus are illustrations of two microbes that are the cause of peptic ulcers and cervical cancer, respectively. Other chronic diseases may be initiated by a virus or bacterial infection. Although not a direct cause, infection with hepatitis B virus and hepatitis C virus can lead to the disease of hepatitis.

Even now, it is more complicated to associate a microorganism with a chronic disease when the organism is the initiator of the disease, but the actual pathology is influenced by additional factors, leading to cases when infection is more prevalent than clinical disease. There are a number of human diseases that fall into this category, still challenging "microbe hunters," including multiple sclerosis and diabetes type 1. The association of viruses with these diseases comes mainly from animal models and immunological studies. In addition, there is increasing evidence to suggest that the herpesvirus, Epstein-Barr virus, is associated with the cause of multiple sclerosis.

As I was writing this, I was reminded of poliovirus–an enteric virus–and the disease of poliomyelitis. Another example where the early tools for identifying and understanding the viral cause was clouded by its complex biology. An important observation in early studies of the paralytic disease leading to the development of the vaccines was that only about 1% of infected individuals experience neurodegeneration caused by the spread of the virus to the central nervous system. Infection of the central nervous system would be responsible for what could be considered the chronic disease of paralysis. Even more relevant might be that decades later some individuals develop post-polio symptoms of muscle weakness and pain thought to be due to a deterioration of motor neurons; poliovirus is no longer present.

I am also reminded of poliovirus as an example of the near miraculous success of vaccines against this virus that have almost led to the eradication of the disease of poliomyelitis. An unstated corollary of Summers' book is the importance and influence of vaccination on the threat of infectious diseases. As of 1980, extensive worldwide vaccination efforts led to the eradication of the disease of smallpox. Now in the 21st century in many countries (although there are definitely exceptions) there is almost no memory about the disease of paralytic poliomyelitis. The pandemics of plague are not a threat. Cases of measles had been rare in many countries, but there has been an increase in infections. Other serious diseases (tetanus, typhoid) are hardly known in many countries, and the importance of vaccination was often not recognized as one of the reasons for the absence of the disease, although that did change with the pandemic of SARS-CoV-2. Instead, there have been protests against vaccination. Perhaps this book will remind us that infectious diseases, known and emerging, remain a threat as cases of polio and measles are reappearing.

There have been protests against vaccination ever since Jenner's vaccine against smallpox, but more recent objections have come from the "antivax" movement—objections to vaccination that seem to have become even more vociferous with the COVID-19 pandemic. While Summers does not delve into the history of vaccines, per se, he does write about the devastation of infectious diseases such as plague, influenza, and others. Understanding microbial diversity and avoidance of microbiological hubris with respect to new and emerging pathogens are essential to our continued success in developing new immunological protections.

After a broad history of diversity in the microbial world and an introduction to the new metagenomic methods coming into widespread use in microbiology, Summers turns to the subject of the human microbiome and especially to the importance of the presence and diversity of the residents of the microbiome. In the past few years this has become a fascinating and explosive field—I believe it is just the beginning. Now the composition of our blood is a normal part of a medical exam—the residents of our microbiome should soon also become a routine part of an exam.

Through most of this book Summers focuses on microorganisms as disease-causing agents. In the last chapter he may be hinting at the subject of his next book – the stable communities of microbes that are found in various physiological niches from deep sea corals to the human digestive track where they play crucial roles in health and disease.

Bill Summers writes that he was strongly influenced by Paul de Kruif's account of the heroes of microbiology, *Microbe Hunters* (1926) and later by *The Microbial World* (Roger Stanier, Michael Doudoroff, and Edward Adelberg, 1957). After reading *Magic Bullets, Miracle Drugs, and Microbiologists*, I am convinced that this book by Bill Summers will be an influence on some young readers to enter the field of microbiology.

Sondra Schlesinger
Professor Emeritus
Department of Molecular Microbiology
Washington University School of Medicine
St. Louis, MO

Preface

This book is the product of three contingencies: my half-century career as a practicing microbiologist, my more recent career as a historian of science, and the recent flood of commentaries on pandemics, probiotics, and the politics thereof.

As a teenager I was strongly influenced as were many of my contemporaries by Paul de Kruif's account of the heroes of microbiology, *Microbe Hunters* (1926). A bit later I encountered a textbook with the title *The Microbial World* (Roger Stanier, Michael Doudoroff, and Edward Adelberg, 1957) which presented microbiology as a comprehensive science and a title that suggests some sort of global unity of microbes. These two ideas, not quite themes, are the two guiding principles in this book: 1) history matters and 2) microbes are, indeed, ubiquitous co-inhabitants of our planet.

I also have drawn upon my understanding of the broad fields of microbiology, infectious diseases, and epidemiology to center this account of the changing views of the microbial world on the decades of the 1960s and 1970s as a paradigm shifting period when both human history with new microbial challenges and scientific progress coincided to change our understanding of this unseen world. Serious "new" diseases such as Legionnaires' disease, Lassa fever, Lyme disease, and a bit later, AIDS, emerged and invaded the public consciousness. At the same time, amazing new genetic approaches such as direct gene analysis by DNA and RNA sequencing provided the new technologies that would change our earlier views of the microbial world.

Prior to these decades, now perhaps seen as infected with microbiological hubris, we were comfortable in our belief that we understood our microbial neighbors, had antibiotics to keep them in their place, and could look forward to a time without major epidemics and other contagious disease disasters. But, to quote that philosopher, Yogi Bera, "It's tough to make predictions, especially about the future."

This book, then, is a historical look at the understanding of our microbial world from its seventeenth century discovery of animalcules to the current appreciation of the diverse microbiomes that are an integral part of our planet, and in some ways a cautionary tale, highlighted by the recent COVID-19 pandemic, to be mindful of the mysteries of microbiology yet to be solved.

Acknowledgments

An earlier version of chapter 4 appeared in Summers WC. 2007. Microbial drug resistance: a historical introduction, p 1–10. In Wax RG, Lewis K, Salyers AA, Taber HW (eds). *Bacterial Resistance to Antimicrobials*. 2nd Ed. CRC Press, Boca Raton, Florida.

I am greatly indebted to many generous mentors, kind colleagues, and generations of Yale students who helped me develop my thoughts on matters treated in this book. I thank Wilma P. Summers, G. Nigel Godson, Sondra Schlessinger, Adam Lauring, and Susan Spath for special help on various aspects of this work. Thanks also to Megan Angelini, my very helpful and diligent editor, for her cheerful and astute advice in bringing this work to publication.

About the Author

William C. Summers is a retired Professor of History of Medicine, of Molecular Biophysics and Biochemistry, and of Therapeutic Radiology and in the Program of History of Science and Medicine at Yale University where he was a faculty member from 1968 until his retirement in 2017. His formal education at the University of Wisconsin included mathematics, molecular biology, and medicine and he received the MD and PhD in 1967. He began his laboratory work on bacteriophages in 1963 and expanded to study animal viruses in the 1980s. In addition to studies on the molecular biology and genetics of cancer and viruses, his scholarship has included history of medicine and science, and the relations between science and the humanities. He has held visiting positions in Sweden, the United Kingdom, Stanford, Columbia University, Hubei Medical College, and the National University of Singapore. His previous historical books include *Félix d'Herelle and the Origins of Molecular Biology*; *Reconceiving the Gene: Seymour Benzer's Adventures in Phage Genetics* (edited); *The Great Manchurian Plague of 1910–1911: The Geopolitics of an Epidemic Disease*; and *The American Phage Group: Founders of Molecular Biology*.

1 Introduction

A SCANDALOUSLY SHORT HISTORY

Microbiome! Who, these days, hasn't heard about "the microbiome"? From TV pitches to "support gut health" to complex explanations in high-end magazines about new understanding of the myriad human-microbe interactions, some apparently essential for our existence as living, breathing organisms. For those of us of a certain age, microbes, aka "germs" were mainly to be casually washed away a few times a day from our grubby little hands before meals and after using the bathroom. How did "microbes" become a central concern in modern life, and what is the history of this recent interest? This question is the theme of this book.

Microbes (a name coined by a French surgeon in the latter part of the nineteenth century) by their nature are not visible to the naked eye, so they were unknown until the mid-seventeenth century when high-powered glass lenses were employed to magnify these tiny objects. The microscope was simply a handy magnifying glass arranged to look at things up-close (1). An amateur scientist from Delft, in the Dutch Republic (at that time science was done by amateurs with regular day jobs because there was no such career path as "scientist.") became very skilled at making lenses and microscopes and recorded his many observations over a period of a half-century (1676–1723). The various objects that Antonie van Leeuwenhoek (1632–1723) observed, he called "little animals" (*animalcules*) because he viewed them as simply tiny versions of the known animals of common experience.

Magic Bullets, Miracle Drugs, and Microbiologists: A History of the Microbiome and Metagenomics,
First Edition. William C. Summers.
© 2024 American Society for Microbiology.

Leeuwenhoek's work became widely known and appreciated, but it took over a century for others to understand just where these animalcules fit into the schemes of life that were being formulated, and what they might be doing in the many environments where they were found. The 19th century was a time of intense expansion in scientific knowledge in many fields; new theories of chemistry, new technologies, new philosophies of nature, and new questions fueled this expansion. The stories of the germ theories of disease and the debates over the living nature of microbes as the agents of fermentation and related everyday processes are by now almost folk tales.

The heroic figures of Louis Pasteur (1822–1895) and Robert Koch (1843–1910) as discoverers of microbes as the causes of many here-to-fore mysterious diseases of both humans and other animals paved the way for the "microbe hunters" of the twentieth century. In the early years of that century, medical scientists (by now "scientist" had become a legitimate job description) sought microbial causes for every known malady, from cancer to mental disorders to heart disease and strokes. These hunters of microbes were remarkably successful. Many serious diseases of humankind turned out to be caused by microbes: tuberculosis, pneumonia, typhoid, cholera, polio, influenza... the list goes on and on (2).

Not only were microbes important in causing diseases, it was found that the animal body had a mechanism to deal with these invading microbes: the immune system. For many diseases, the body could react, over time, and develop powerful defenses against a later infection with the same or a related microbe. In a way, the body learned from its first encounter. The way this immunological learning works has taken nearly a century to unravel. But even before this process was completely understood, the phenomenon of immunity was quickly exploited to devise preventative measures. A deliberate infection (under mild conditions, it was hoped) could be used to induce this immunological protection against later, more dangerous, natural infections. Historically, smallpox was the human scourge most widely prevented by this inoculation procedure. Later, in 1796 Edward Jenner (1749–1823) introduced a novel variation on smallpox inoculation when he recognized that a related but benign infection with material from animals with cowpox induced immunity to smallpox just as well as a mild case of smallpox itself. This kind of immunization became known as "vaccination," a name derived from *vacca*, Latin for a cow.

In addition to the development of immunizations against many microbes, from the early years of the twentieth century the pharmaceutical chemists were developing drugs to suppress microbial diseases. Some of these drugs were quite successful. One such drug made from arsenic and called Salvarsan became the first really useful treatment for the dreaded syphilis. The real "breakthrough" and the real start of this account came in the mid-decades of the twentieth century with the discovery of several drugs with broad use against many different microbes, drugs such as the sulfa drugs. These drugs were based on the known metabolic reactions shared by many microbes but not their animal hosts. But the main event turned out to be the recognition that various microbes compete with each other in the environment and have evolved toxins to destroy their competitors. In 1928, famously, Alexander Fleming (1881–1955) in London observed a case of this microbial antagonism on some of his accidentally contaminated culture plates of bacteria that had been invaded by a mold. The mold, *Penicillium notatum* (now called *P. chrysogenum*), a common bread mold, produced a soluble substance, a chemical that Fleming partially isolated and showed would inhibit the growth of certain disease-causing bacteria. He named this substance penicillin. Because it was from a biological source, the mold, and because it was useful to inhibit other organisms, penicillin became known as an "antibiotic" to be distinguished from the antimicrobial agents produced in the lab by chemists (3). For various technical reasons, it took nearly 15 years to produce and purify enough penicillin to show that it was clinically useful to treat microbial infections. Fleming's discovery, expanded on by others, about microbial antagonism led the former microbe hunters to become antibiotic hunters. Many new and useful antibiotics were discovered, tested, and brought into routine clinical use. There was hardly an infection known to medicine for which a useful antibiotic could not be found. Thus begins our story.

MICROBIAL HUBRIS

What microbes do we know about, and, how do we know them? Two problems confronted the early microbe hunters: what methods are available to hunt microbes, and how should we classify, describe, and catalog our quarry. The first question is primarily one of laboratory technology, the second is practical as well as philosophical.

The microbiologists of the nineteenth century employed several basic techniques: inoculation of suspect material into various experimental animals, usually guinea pigs or mice and the watch for changes in health, appearance, or behavior. The affected animal was then dissected to look for both macroscopic and microscopic traces of the suspect microbes. A more focused method was to smear suspect material on a surface of some substance that was both free of known microbes, i.e., was "sterile" and was believed to support the growth of a microbe if it was present. These sterile culture surfaces were sometimes gelatin or agar surfaces that had been sterilized in an oven, or sometimes just the cut surface of a potato. The gelatin or agar was supplemented with sugars, meat broth, or other "nutrients" that the microbe hunter believed (or guessed) was needed by the microbe to grow and flourish on the surface. If the conditions were right, a microbe that landed on the surface and started to grow would eventually form a visible "colony" of all the millions progeny of the original microbe, a colony of identical descendants, a so-called pure culture. Some of these microbes could then be observed under the microscope, tested in other ways, and eventually characterized and identified as either a known type or as a new, unknown, isolate. For a long time, basically up until the molecular revolutions of the 1970s, the main way microbes were described and characterized was by their size, shape, and appearance of their colonies, as well as their chemical properties as identified by their staining with various dyes, their growth or non-growth on culture media of varying nutritional composition, or by their reaction with known antibodies (the molecules of the immune system that had been stimulated by specific known microbes to produce these identifying tools). With these rather simple and universally available technologies, the microbe hunters could be reasonably confident that microbes found in nature in one part of the world or by one laboratory could be compared with microbes found elsewhere.

With descriptions and characterizations of microbes in hand, the microbe hunters had to decide how to organize their collection. In the very early days in the nineteenth century, they relied on microscopic size and shape plus visible appearances of the colonies on various culture setups. As chemical and cultural technologies advanced and the relationship between the conventional classifications of Carl Linnaeus (1707–1778) and the evolutionary principles of Charles Darwin (1809–1882) were recognized, new microbial taxonomic systems developed. By about 1920, consensus was developing around certain principles, and several groups of microbiologists came together to produce what very soon became the standard authority on bacterial classification (viruses and non-bacterial parasites were more complex and controversial). Under the leadership of David H. Bergey (1860–1937), a professor at the University of Pennsylvania, *Bergey's Manual of Determinative Bacteriology* was first published in 1923 (4). The very term "determinative" indicated that this was a source for information to determine what kind of bacterium one had just isolated. This manual, and its many subsequent updates, provide microbiologists with a consensus classification system, approved methods for characterizing bacterial isolates, and standardizations to insure inter-laboratory comparisons and identifications. *Bergey's Manual* (for short) integrates bacteria into the general biological classification scheme derived from the eighteenth-century views of Linnaeus, the so-called binomial classification that assigns hierarchical groupings down to the final categories of genus and species. Species, is, of course, considered the individual identification that defines the biological specificity. It is worthwhile to note, however, that the species concept is not without its critics, and modern biology struggles to come to grips with the species concept as more and more examples are found that don't neatly fit. In microbiology, virus classification in particular, seems resistant to this traditional view.

THE MICROBIAL WORLD

All in all, however, microbiologists during the middle decades of the twentieth century could culture away, finding, and naming organisms from their environment and their patients, following Bergey and discovering ever more examples that fit nicely into the niches of Bergey's groups and flow charts. Identification of new germs or recognition of known germs as causal agents of disease or perhaps even useful processes (think wine and cheese) was one thing, finding treatments or modifiers of microbial growth and pathogenicity was another. Although the French chemist Pasteur was famous for his pioneering work on various immunological approaches, introducing many useful inoculations (i.e., vaccines) and the serum from animals (and even humans) which contained the antibodies induced by these vaccines to treat or prevent some microbial infections, science was usually unable to find effective treatments for some of the most troublesome infections. The German chemist, Paul Ehrlich (1854–1915) took another, more direct, approach. Ehrlich noted that the stains used to characterize and visualize them were often the same compounds that could kill these microbes. The specificity of the staining reactions coupled with their lethal action suggested to Ehrlich that it might be possible to find specific chemicals that could treat specific infections. His first success in 1910 was the development of a compound (now known to be a mixture of related compounds) that contained arsenic called arsphenamine, also known as Salvarsan or compound 606, that could successfully treat both syphilis and African sleeping sickness (African trypanosomiasis). Because this chemical was both effective and highly specific, Ehrlich employed the metaphor of the "magic bullet," a phrase that has become part of our vernacular discourse (5). Salvarsan became the first example of what became known as chemotherapy. It was several decades, however, before the next magic bullet was produced in the form of the sulfa drugs (introduced in 1935). These drugs would prove to be the first in a long line of small molecule drugs that effectively targeted diverse bacterial species.

After Fleming's discovery of penicillin and the recognition of the microbial complexity of ecosystems such as soil, the discovery and isolation of several other antibiotics soon followed. Streptomycin, isolated in 1943 from the soil microbe *Streptomyces griseus*, was the first such antibiotic that was broadly effective and showed only rare side effects in humans. Others were discovered rapidly based on the same principles: actinomycin (1940), cephalosporins (1945), chlortetracycline (1945), and chloramphenicol (1947). Immediately following WWII when penicillin was introduced and sulfas were used to control infections, both penicillin and streptomycin were hailed as "wonder drugs" or "miracle drugs" and flowed like water across the medical landscape of the developed world. Infections that were previously feared as a death sentence, pneumonia in the elderly, tuberculosis, sepsis, typhoid fever, gonorrhea, and syphilis, all fell victim to the curative powers of the new miracle drugs being produced by the microbe hunters and their friends, the pharmaceutical chemists.

The new antibiotics were so successful and apparently so innocuous that they became the "go-to" answer for every medical encounter. As we now know, this selective pressure (*pace* Darwin) led to microbial population shifts and the massive problems with drug resistance that are now commonplace. Still, acute infectious diseases seemed controlled, something to be treated with the ever-new generations of antibiotics being discovered... up to a point. Between 1935 and 1968, 12 new classes of antibiotics were introduced but between 1969 and 2003, only two more (6).

The clinician, however, when confronted with a potential infection as well as a patient clamoring for one of the new "wonder drugs" simply pulled out the syringe and provided a dose of penicillin. The patient got well but whether or not it was due to the penicillin was never an issue... it was another score for the power of antibiotics. Antibiotics, it seemed to many, heralded the end to the scourges of infectious diseases of life on Earth, animal and plant alike. Soon, however, two biological realities changed all that: microbial diversity and genetic variation.

With the widespread use of antibiotics, as Darwin predicted, selection for resistant variants became the norm. Secondly, microbial diversity, long considered a naturalists hobby, revealed the vast array and abundance of microbes just waiting for ecological opportunities to show themselves in new, sometimes troublesome environments. These developments coincided with a rather abrupt change in microbiological technologies from morphological and physiological methods to genetic methods, all occurring in the few decades just after mid-twentieth century.

Microbiologists, both in the lab and in the clinic, were justly proud of their early twentieth century successes at identifying the causal microbes underlying many of the prevalent animal and plant diseases. The discovery of antibiotics also was an accomplishment that led to optimism for a future free of sickness and death from major infections. Indeed, as several noted physicians have recently recalled, microbiology in the 1950s and early 1960s was marked by a kind of scientific hubris: we conquered the microbial world. This attitude, perhaps justifiably, led to a general malaise in the field of medicine known as "infectious diseases." This book traces some of the events, both in the lab and in the clinic, that upended this optimistic outlook in the 1960s and 1970s. With the emergence of unknown new infectious diseases and the recognition of our ignorance of the vast majority of unknown members of the microbial world, we again lost our innocence as scientists and clinicians and now grapple with the new concept of microbiomes and the powerful new technologies of metagenomics.

NOTES AND REFERENCES ───────────────────

1. New understanding of the science of optics in the early seventeenth century produced both microscopes as well as telescopes, another device with lenses arranged to look at things far away.
2. The meaning of "causation" is a traditional problem in medicine. In this context of infectious diseases, it most frequently has the meaning of "necessary causation," that is, the specific microbe must be (or has been) present for the disease to occur. The microbe may or may not meet the criterion of "sufficient causation," that is, its presence alone is sufficient to (always) cause the disease.
3. **Bentley R, Bennett JW**. 2003. What is an antibiotic? Revisited. *Adv Appl Microbiol* **52**:303–331.
4. **Bergey DH**. 1923. *Bergeys's Manual of Determinative Bacteriology*. Williams and Wilkins, Baltimore, Maryland.
5. The magic bullet reference was introduced by Paul Ehrlich in 1907. He used the German term *Zauberkugel*, and the English translation "magic bullet" in The Harben Lectures at London. This term has its origins in the Teutonic folkloric character of the *Freischütz*, a marksman who, by a contract with Satan, has obtained magic bullets destined to always hit the intended target. Of course, there is a hidden cost: the last bullet is at the disposal of the Devil himself. The famous opera *Der Freischütz* (1821) by Carl Maria von Weber treats this traditional theme.
6. **Conly J, Johnston B**. 2005. Where are all the new antibiotics? The new antibiotic paradox. *Can J Infect Dis Med Microbiol* **16**:159–160.

2 Microbe Hunting before DNA

The hunting of microbes, of course, relies on knowing the existence of our quarry. Before the middle of the 1600s microbes were unknown, that is, while they certainly existed, humankind was blissfully unaware of them. In the early part of the 1600s (the seventeenth century), in Europe the science of optics was making rapid progress... good magnifying lenses were being made and their properties were analyzed. The science of optics spilled over even into art and the science of perspective became required study, especially in the Dutch and Italian schools of painting (1). Two new instruments were devised that were to revolutionize science and philosophy: the telescope and the microscope. Both instruments provided humans with new experiences with previously invisible aspects of our universe, the very large and distant, and the very small and up close. We will concentrate on the tiny new world of microbes, but the distant world of the cosmos was equally disruptive to existing beliefs about the world, nature, god, and human beings' place in it.

The first microscopes were not the elaborate and beautiful instruments now familiar to most high school biology students. They were simply devices that allowed a glass lens of rather high magnifying power to be focused very close to some object to be examined. Often a good lens was simply a well-made spherical drop of glass which, because it was small, had a high curvature and hence provided a high degree of magnification. The single glass lens was mounted between two brass plates that also had a few little screws with a mounting pin upon which to impale the object to be studied in the right place near the lens. An insect, a bit of

Magic Bullets, Miracle Drugs, and Microbiologists: A History of the Microbiome and Metagenomics,
First Edition. William C. Summers.
© 2024 American Society for Microbiology.

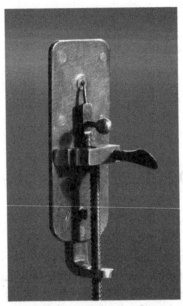

FIGURE 2.1 *Replica of one of Leeuwenhoek's many single lens microscopes. The lens is a spherical glass bead held between two plates of brass. The sample is impaled on the point of the adjustable needle very close to the lens. The observer (if sufficiently near-sighted) holds the device near to one eye to observe the magnified object through the lens. The screw adjustments are to enable positioning the sample so it is in focus relative to the focal length of the lens (13). Photo credit: Jeroen Rouwkema, under license CC BY-SA 3.0.*

plant material, or a droplet of liquid were some of the objects that interested the very first microbe hunters. Figure 2.1 shows one of these first microscopes.

The individual responsible for the development of these initial microscopes as well as extensive application of such instruments to an extensive range of natural objects was a Dutch businessman living in Delft in the mid to late seventeenth century and on into the eighteenth century. Antonie van Leeuwenhoek made his living trading in fabrics and became recognized as a respected citizen of Delft (2). He apparently became curious about this new instrument from a slightly earlier pioneering work on the minute unseen world, a compendium of observations by the Englishman Robert Hooke (1635–1703) in his magnum opus, *Micrographia* (3). Leeuwenhoek became proficient at making fine lenses, and he was believed to have made on the order of 500 different microscopes, some of which are still in existence.

Leuwenhoek was a self-educated scientist, an amateur in the best sense of the word, and, upon the urging of a local, but famous physician, Regnier de Graaf (1641–1673), sent some of his microscopic observations to the Secretary of the

Royal Society of London. The secretary, Henry Oldenburg (1619–1677), was so impressed with these new findings that he had Leeuwenhoek's Dutch work translated into English or Latin (the languages of the *Philosophical Transactions of the Royal Society* at that time) and published. This correspondence extended over a period of 50 years and made Leeuwenhoek famous as the "Father of Microbiology" (4).

Leeuwenhoek's observations on diverse samples from sources such as pond water, melting snow, saliva, and food stuffs suggested to him that what he called "small animals" (*animalcules*) were ubiquitous in nature. In addition to describing and providing drawings of these objects, he recognized certain regularities and started to classify them, mostly by size and shape. He had no way to know that these objects were alive in the biological sense, but because some were believed to be motile (apparently swimming), by analogy with large, macroscopic, organisms of everyday experience, he felt justified in thinking of them as simply small examples of the animal kingdom.

Historians date Leeuwenhoek's landmark report, one that has been seen as the beginning of the science of microbiology, to his letter to the Royal Society of 9 October 1676. In this letter he described what came to be known as "infusoria" (the microbes found in infusions, water in which plant material has been soaked, for example, tea). Leeuwenhoek described the little organisms he observed in a drop of pond water, first noted in the summer of 1674 and then extended to diverse infusions, both natural and artificial. He famously enthused that the observations of the many little animals in a drop of water was "among all the marvels that I have discovered in nature, the most marvelous of all" (5). His descriptions and drawings are sufficiently precise and detailed to allow the modern microscopist to be quite sure of the various organisms that he was encountering (Figure 2.2).

Although some others had difficulty confirming Leeuwenhoek's observations (usually because they had inferior instruments), others were able to confirm and extend his observations over the ensuing years. While the new microscope was employed in study of the invisible structures of many things, and led to astounding advances in understanding of anatomy, "at the microscopic level," so to speak, the "little animals" remained a relatively ignored finding for nearly a century.

At this point it is useful to introduce a little philosophy. . . just for a moment. Probably the earliest humans, when thinking about the world they inhabited, took to the business of classifying or categorizing their observations. Plants versus animals, humans versus plants, edible versus inedible, and so on. By the time of Aristotle (384–322 BCE) this apparently natural pattern of thought caught the attention of serious thinkers. Philosophers since then, right down to the present, have explored our ways of classifying things. One key principle is based on the observable qualities of objects: size, color, use, origin, and so on. Early on, it was

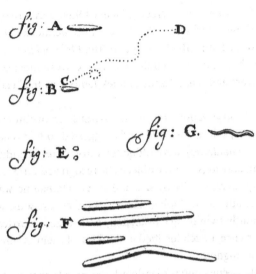

FIGURE 2.2 *First image of a microbiome. Leeuwenhoek's drawing of animalcules from the human mouth. Letter 39, September 1683 (14).*

more or less agreed that some qualities could be considered as primary, while others were secondary. Primary qualities were those observables that could not be absent for the object to still be recognized as that object. Further, primary qualities were essential to the identity or "true nature" of the object. It has become common to speak of classifications based primarily on these natural primary qualities as forming groupings called "natural kinds;" the category "fruits" represent what many would call a natural kind, whereas a category of "all red objects" would not. Just for the record, however, professional philosophers still argue about the whole notion of natural kinds. "Despite its long history and intuitive appeal, the conception of species as natural kinds is difficult to sustain while also maintaining a traditional view of what a natural kind requires: a set of intrinsic natural properties that are individually necessary and jointly sufficient for a particular to be a member of the kind" (6).

Since classification and recognition of biological diversity of microbes requires some sort of classification, it is good to keep in mind just what the process of classification entails.

Another relevant legacy of the past that has had an important role in the history of microbiology also stems from Aristotle: the *scala naturae*, Latin for the Ladder of Nature. In trying to devise a comprehensive and rational classification scheme, Aristotle devised not only categories but also a linear ordering of all of nature. While it may seem a bit biased in favor of Aristotle's place in nature, he put minerals (earth) at the bottom (less valued?) and humans near the top of the ladder, second only to spirits and gods. His scheme was based on a rather nested series of primary qualities, such as movement, reproduction, cognitive qualities, and so on. This attractive scheme, later called the "Great Chain of Being," was, as you might imagine, attractive to the Western religious thinkers of the medieval period, and was very widely incorporated into Christian theology since the medieval period when Aristotelian thought was being recovered in Europe. In its medieval form, the rungs of the ladder placed minerals at the bottom with the next rung being the fungi, thought to be a "lower" form of the plant category. "For centuries the 'great chain of being' held a central place in Western thought. This view saw the Universe as ordered in a linear sequence starting from the inanimate world of rocks. Plants came next, then animals, men, angels and, finally, God. It was very detailed with, for example, a ranking of human races; humans themselves ranked above apes above reptiles above amphibians above fish. This view even predicted a world of invisible life in between the inanimate and the visible, living world, long before Antonie van Leeuwenhoek's discoveries. Although advocates of evolution may have stripped it of its supernatural summit, this view is with us still" (7).

There was a big gap between minerals and plants... until Leeuwenhoek's observations on the "little animals." But this gap was only recognized retrospectively in the nineteenth century in the search for "missing links" in the geological record prompted by attacks on Darwin's ideas such as descent with modification. At last, however, just perhaps... the chain of being between the dead objects (earth) and the living objects (plants) could be linked up by the living, but primitive forms of life observed with the new tool, the microscope.

At this point we encounter one of those interesting characters in science whose talents seem uniquely fitted to his time and its problems. Christian Gottfried Ehrenberg (1795–1876) was a German naturalist who was lucky enough to find a life-long mentor in one of the most famous scientists of his day, Alexander von Humboldt (1769–1859). In contrast to the modest station in life of Leeuwenhoek, Ehrenberg was born into a comfortable upper middle-class family near Leipzig and received a broad education first in theology and then in medicine and natural science. His doctoral dissertation was on fungi. After an extensive scientific expedition through the Middle East during the years from 1820 to 1825 collecting many specimens and conducting an influential study of corals of the Red Sea, Ehrenberg returned to the University of Berlin as professor of medicine in 1827. Two years later he joined Humboldt for another expedition, this time through eastern Russia to the Chinese frontier. At this point he began his study in earnest on microscopic organisms, a relatively ignored field until then. Ehrenberg had been a skeptic of the biological implications of the Great Chain of Being notion from an early age and saw the study of microscopic forms as directly relevant to this biological and theological problem. An individual reputed to be both capable and charismatic, he became an effective advocate for science and microscopy in particular. In what many scholars view as the magnum opus among an extensive output of published work, *Die Infusionsthierchen als vollkommene* [The infusion animals as perfect] (1838), Ehrenberg described many forms and suggested that far from being primitive, perhaps intermediate forms, these infusoria were complete little animals, basically miniature forms of the larger animals we all knew. His diagrams show these organisms with gullets, stomachs, and appendages, purported to refute their status as the intermediate "links" between minerals and fungi as then postulated in the Great Chain dogma (8). (Figure 2.3).

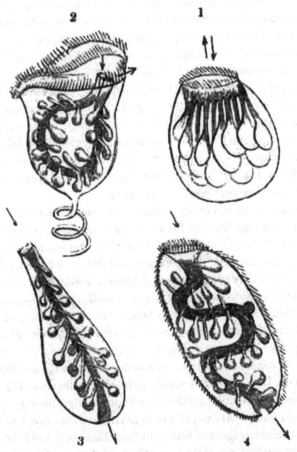

FIGURE 2.3 *Image of some infusoria as drawn by von Ehrenberg in Die Infusoria (1838) (15). The arrows indicate von Ehrenberg's interpretation of the feeding mode of these organisms. Note the presence of structures analogous to the digestive systems of higher macroscopic animals.*

While Ehrenberg's work on the anatomy and physiology of infusoria was challenged and subsequently discarded, his work on categorizing, classifying, and naming the myriad of organisms that he observed over a 30-year period served to establish the study of microbes as an interesting and fruitful aspect of nineteenth century science, and his massive and beautifully illustrated tome on *Infusoria*, even if somewhat fanciful, is still admired by historians of science.

The nineteenth century saw major improvements in microscope design, versatility, and optical qualities, and the pioneering work of Ehrenberg and a small group of like-minded scientists set others to work exploring the microbial world in earnest. Ferdinand Cohn (1828–1898), a German Jew born in Breslau (now Wrocław in Poland), became the leader in the new science of microbes under the influence of his teacher, Ehrenberg, when he studied in Berlin (9). Cohn returned to his native city to become professor of botany at the University of Breslau in 1871. There he established an institute of plant physiology, a scientific journal, *Beiträge zur Biologie der Pflanzen* [Contributions to the biology of plants], devoted to the publication of work from his institute, and set up a marine aquarium to provide study materials far from the oceans. The status of the microscopic organisms as either plants or animals was a hotly debated issue in the nineteenth century as the drive to categorize and classify them was strong. The small size of these objects necessitated the interposition of a sophisticated instrument, the microscope, between nature and naturalist. When it came to listing the properties or observable qualities that could provide means to identify and classify them, size, shape, color, motility, and sometimes mode of cell division, were about the only ways to describe them as objects. Later, of course, some functional qualities such as natural habitats, disease associations, and the like were incorporated as primary qualities. Cohn was particularly concerned with color (green) which might help decide whether an organism might be classified as a plant (or non-plant). Many of his studies were on microscopic algae. By the latter half of the nineteenth century, the cell theory of Matthias Schleiden (1804–1881) and Theodor Schwann (1810–1882) had taken hold and was adopted by microscopists who studied the little animals. Cohn defined the subgroup of microscopic organisms that came to be called bacteria as "chlorophyll-free cells of spherical, oblong, or cylindrical form, sometimes twisted or bent, which multiply exclusively by transverse division and occur either isolated or in cell families" (10). Cohn, director of an institute of plant physiology, would also go on to adopt physiological characters as primary qualities that could be used to classify microbes. As laboratory growth methods ("culturing") became routine, the specific nutritional requirements of various microbes became recognized as primary qualities, useful in classification schemes.

Of course, no survey of microbial classification and discovery would be complete without mention of the work of Louis Pasteur in France and Robert Koch in Germany. While the discoveries of Pasteur and Koch related to causing infectious diseases is well-known as is Pasteur's researches on immunizations, for our story we can turn to their somewhat lesser known, but crucial contributions to the basic science of microbiology, namely their work to establish the concept of a pure culture. This work has been crucial to understanding both the identification and classification of microbes, but also their genetic analysis in more recent times.

Microbes exist in nature as complex mixtures for the most part. The infusoria of Leeuwenhoek and Ehrenberg derived from pond water, various "teas," and animal fluids show diverse forms present in varying proportions. As Pasteur and Koch, separately and independently, realized from their work on the role microbes played in fermentations, infectious diseases, and industrial processes, and from the growing knowledge of specific growth requirements of certain microbes, a method to separate and propagate a single kind of microbe for detailed study of its unique properties would be needed. They needed "pure cultures" of a single organism (11).

Pasteur approached this problem by growing his cultures in liquid medium ("broth") which had the required nutrients for growth of the microbe of interest. After the microbes had a chance to grow and multiply extensively in the flask, he took a tiny sample and transferred it to a fresh flask of sterile broth and let it grow up again, after several transfers and re-growth, he noted that he had produced a homogeneous population of microbes, his "pure culture." Pasteur's liquid serial dilution technique, as you might have realized, was a form of Darwinian selection, where the most efficiently growing microbe would be the winner in the end. If one knew enough about the growth conditions that favored the organism you aimed to purify, all went well. If not, however, you could end up with the unwitting selection of an initially minor component of your original sample. Also, this method gave very little information about the microbial diversity of your samples.

Koch on the other hand, grew his bacteria on the surfaces of solids that provided nutrition to the microbes of interest. His method, where he first used the sterile surface of freshly cut potatoes, had the advantage of spatially locating a given microbe as well with all of its progeny at a specific place on the surface (where it happened to settle down after he spread the original sample on the surface. The progeny would eventually become so numerous that the "colony" of them (all related to the original single founder microbe) became visible to the naked eye and could be sampled with a sterile needle for inoculation into broth culture or onto more solid surfaces for study. One of Koch's students, Richard Petri (1852–1921), made improvements after the potatoes had been abandoned and replaced with broth solidified with gelatin in a flat dish (now bearing his name); at the suggestion of Fanny Hesse (1850–1934), an American woman working as a volunteer assistant in Koch's laboratory, they replaced the

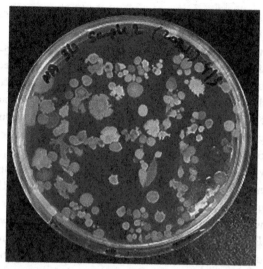

FIGURE 2.4 *Petri dish with nutrient broth solidified by the addition of agar as a gelling agent (16). The medium was sterilized and then after cooling and solidifying a sample of sputum was spread on the surface and the dish incubated overnight at body temperature. The spots are made up of homogeneous populations of different kinds of bacteria. Each spot is a colony of microbes that grew up from a single cell that happened to settle on that place on the surface of the solid agar medium. The size, shape, and color of each colony is a useful property for identification of the species of bacteria in that colony. © 2021 Kaashyap M, Cohen M, Mantri N, under license CC BY 4.0.*

gelatin with the plant gelling agent, agar, a method still in use today (12). Koch's solid medium culture had the distinct advantage of allowing the diversity of the microbial population to be observed and to permit the study of rare forms that can be recognized in complex mixtures (Figure 2.4).

Pure culture methods, coupled with advances in biochemistry and physiology, and the introduction of chemical staining of microscopic specimens in the late nineteenth and early twentieth century led to the growth of bacteriology into a full-fledged discipline with its own society (the Society of American Bacteriologists, later the American Society for Microbiology, founded in 1899), a proliferation of scientific journals, and the founding of academic departments. The question of bacteriological classification and recognition of the diversity of species remained, however. Classification schemes served several functions: standardization of names, agreement on the properties that define a given genus and species, procedures for recognition of newly discovered organisms, and pathways to expand and revise the classification schemes themselves. One goal that was late in being applied to microbes, however, was the notion of biological relatedness. For large organisms, say cats, we have a natural feeling that house cats, lions, and tigers all belong to some sort of natural kind. This natural feeling was initially absent in bacteriology: all tiny round bacteria did not elicit the same confidence that they should be grouped as a natural kind.

Charles Darwin contributed greatly to our thinking about classification in an indirect way. His notion of "descent with modification" suggested that primary qualities of an organism should be heritable from a common ancestor, and thus classification schemes should be congruent with evolutionary relationships. This profound concept introduced genetics into biological classification. The problem for microbes, however, was that their genetic structure and behaviors were poorly understood until the second half of the twentieth century. Indeed, serious geneticists disputed whether, or not, bacteria even had genes as late as the 1940s.

The confluence of biochemistry, genetics, and bacteriology in the middle decades of the twentieth century under the guise of molecular genetics provided hints for the way forward in microbial classification schemes and new ways to hunt for new microbes. First, studies of microbial nutrition and metabolism spurred on by industrial uses of microbes and the realization that bacterial metabolism could be a target for drugs aimed at infectious diseases provided new markers of bacterial classification. Soon these markers were shown, in many cases, to be controlled by the genes of the organism. Microbial genetics flourished and a robust new way to classify bacteria emerged.

We have already met the "bible" of bacterial classification, *Bergey's Manual of Determinative Bacteriology*, first published in 1923. From the first edition to the current edition (five volumes, 2005–2012), there has been a steady increase in the emphasis on genetics and biochemistry as the basis for bacterial identification and classification. The Darwinian theme of classification reflecting evolution has become ever stronger.

As we will soon see, the new tools of genetics led to the explosion in our understanding of microbial diversity that is the theme of this book.

NOTES AND REFERENCES

1. **Dijksterhuis EJ.** 1970. *Simon Stevin: Science in the Netherlands around 1600.* Martinus Nijhoff, The Hague, Netherlands.
2. **Snyder LJ.** 2015. *Eye of the Beholder: Johannes Vermeer, Antoni van Leeuwenhoek, and the Reinvention of Seeing.* WW Norton & Company, New York, New York.
3. **Hooke R.** 1665. *Micrographia: or Some Physiological Descriptions of Minute Bodies Made by Magnifying Glasses. With Observations and Inquiries.* Jo. Martyn, and Ja. Allestry, printers to the Royal Society, London, UK. A modern reprint edition was published in New York by Dover, 1961.
4. **Dobell C.** 1932. *Antony Van Leeuwenhoek and his "Little Animals"; Being some Account of the Father of Protozoology and Bacteriology and his Multifarious Discoveries in these Disciplines; Collected, Translated, and Edited from his Printed Works, Unpublished Manuscripts, and Contemporary Record, Published on the 300th Anniversary of his Birth.* Bale, London, UK. See also: **Fournier M.** 1996. *The Fabric of Life: Microscopy in the Seventeenth Century.* Johns Hopkins University Press, Baltimore, Maryland. pp. 167–184.

5. van **Leeuwenhoek A.** 1677. Concerning little animals. *Philos Trans R Soc* (12):821–833.
6. **Bird A, Tobin E.** 2022. Natural Kinds. *In* **Zalta EN** (ed), *Stanford Encyclopedia of Philosophy*, Spring 2022 Edition. https://plato.stanford.edu/archives/spr2022/entries/natural-kinds/.
7. **Nee S.** 2005. The great chain of being. *Nature* **435**:429.
8. **Grote M.** 2022. Microbes before microbiology: Christian Gottfried Ehrenberg and Berlin's infusoria. *Endeavour* **46**:100815.
9. **Cohn F.** 1901. *Ferdinand Cohn: Blätter der Erinnerung;* compiled by his wife Pauline Cohn; with contributions by F. Rosen. 2nd Ed. J.U. Kern's Verlag, Breslau, Germany (Wrocław, Poland).
10. **Cohn F.** 1873. Memoirs: researches on bacteria. *J Cell Sci* **2**:156–163.
11. Their work would also address a vexing problem of late nineteenth century bacteriology; how do bacteria fit into the accepted biological classification scheme of "genus and species" devised by Carl Linnaeus in the eighteenth century? Some microscopists believed there was only one (or a very few) true species of bacteria and that the diversity of forms seen under the microscope represented variation due to growth conditions, developmental stages, and so on. This view was called "pleomorphism" because one bacterial species could exhibit many forms. In opposition to this view was the hypothesis of "monomorphism," the view that each bacterial species had but one form and that the observed diversity of forms meant that, indeed, there were a diversity of bacterial species. Needless to say, except in rare cases, monomorphism prevails in current thinking.
12. **Petri RJ.** 1887. Eine kleine modification des Koch'schen plattenverfahrens. *Centralbl Bakteriol Parasite* **1**:270–280.
13. **Rouwkema J.** 4 March 2008. A replica of a microscope by Van Leeuwenhoek. https://en.wikipedia. org/wiki/Antonie_van_Leeuwenhoek#/media/File:Leeuwenhoek_Microscope.png. Accessed 15 December 2023. License: https://creativecommons.org/licenses/by-sa/3.0/.
14. van **Leeuwenhoek A.** 1695. *Arcana Naturae Detecta*, p 42. Henricum a Krooneveld, Delphis Batavorum.
15. **Ehrenberg GC.** 1838. *Die Infusionsthierchen als vollkommene Organismen.* 2 vols., L. Voss, Leipzig.
16. **Kaashyap M, Cohen M, Mantri N.** 2021. Microbial diversity and characteristics of kombucha as revealed by metagenomic and physicochemical analysis. *Nutrients* **13**:4446.

3 Magic Bullets and Miracle Drugs

MAGIC BULLETS: SULFA, TYROTHRICIN, AND STREPTOMYCIN

While the pioneering work of Paul Ehrlich at the beginning of the twentieth century on arsenic compounds as a specific treatment for syphilis demonstrated that the principle of treating infectious diseases by killing off the offending microbes shown to cause them was sound, progress in finding such microbe-specific drugs was slow to non-existent. Many substances killed microbes but it was nearly impossible to find ones that were specific to the microbe while sparing the host. This differential toxicity was hard to come by.

The small community of scientists who took on this challenge followed one of two approaches: a naturalist's approach to search for natural examples based on specific microbial antagonisms, or a biochemical approach based on understanding the physiological differences between microbe and host that might be exploited to design specific therapies aimed at microbe-specific physiology. In the 1930s both approaches finally had some success. In Germany scientists at the Bayer laboratories reasoned that microbe-specific chemical dyes must bind and interact with some components of the organisms and possibly exert lethal effects. Since Bayer was a powerhouse in dye chemistry (they discovered ways to make synthetic indigo, the original dye for blue jeans as well as military uniforms) they were well-positioned to exploit this approach. In 1932, after a long search for such dyes, researchers led by Gerhard Domagk (1895–1964) found one in the family of their original aniline dyes from the late nineteenth century, a sulfonamide they

Magic Bullets, Miracle Drugs, and Microbiologists: A History of the Microbiome and Metagenomics,
First Edition. William C. Summers.

called "Prontosil" (1). It was effective against a range of troublesome streptococcal infections in both laboratory animals and human patients. Interestingly, Prontosil has no effect on bacteria in laboratory cultures, but as was later found, its effects on bacterial infections depends on its conversion in the animal host body into an active form; in modern terminology, it is a pro-drug. This first sulfa drug led to the development of a wide variety of sulfas, fulfilling the dream of Paul Ehrlich of finding "magic bullets" in the chemistry lab.

At the same time, biologists interested in what we now think of as ecology were starting to look to natural environments for microbes that might be sources of antibacterial substances. While not directly involved in antibacterial drug hunting, Sergei Winogradsky (1856–1953), a Ukrainian-born Russian biologist of the late nineteenth century is certainly one of the earliest theorists of antibiotic biology. Winogradsky was educated in St. Petersburg and Strasbourg and had an international career in Switzerland, Russia, and France. Although he formally mentored only one student, he had a major influence on microbiology through his work on microbial ecology and metabolism, especially in complex ecosystems, especially soil microbial populations. He is mostly remembered today for the "Winogradsky column," still used today as an educational demonstration of complex soil ecology (see Figure 3.1) (2). It was Winogradsky's work combining individualized

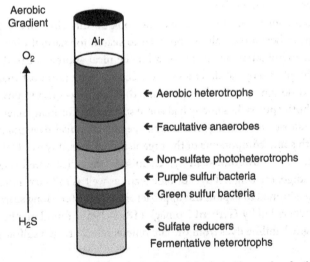

FIGURE 3.1 *Schematic Winogradsky column; the column is loaded with samples of soil, cellulose, and water, and sealed to the atmosphere. Subsequently, a gradient of oxygen is established from the air at the top to anaerobic conditions at the bottom. Microbial growth appears in layers populated by microbes with different growth requirements.*

microbial biosynthesis with mixed microbial ecosystems that provided guidance to the next generation of microbiologists who sought antibacterial chemotherapies.

One of these scientists was René Dubos (1901–1982), a French-American polymath now remembered for his environmental activism, but who, in his early research work pioneered the hunt for antibiotics (3). In 1927 Dubos joined Oswald Avery (1877–1955) at the Rockefeller Institute and joined in the search for enzymes that might dissolve the coating of pneumonia bacteria, a coating that Avery had shown was responsible for the disease-causing behavior of these microbes. Dubos and Avery saw the analogy of this process of bacterial degradation to that which happens in the soil where bacteria degrade similar organic matter on the forest floor. Following Winogradsky, they sought and eventually found such an enzyme from a soil microbe (4). Although it was not the hoped-for magic bullet for pneumonia, it did put Dubos on the right track and by 1939 he and his colleague Rollin Hotchkiss (1911–2004) went on to discover the first true antibiotics, tyrothricin and gramicidin, medicines still in limited use today. In his very first account of this discovery, Dubos sets out his theory of antibiotic hunting: "It is based on the assumption that all organic matter added to the soil eventually undergoes decomposition through the agency of microorganisms" (5).

Another follower of Winogradsky in the U.S. was Selman Waksman (1888–1973) working at Rutgers University in New Jersey. Waksman was a soil microbiologist in the tradition of Winogradsky and had a particular interest in marine microbes and their capacity to degrade organic material in both sea water and marine muds. Waksman, like Avery and Dubos, saw the natural degradation reactions by microbes in the soil as an interesting and fruitful place to study bacterial antagonisms (6). But as late as 1937, Waksman was only theorizing about the potential applications of this field to antimicrobial therapy (7). In a retrospective account of Waksman's conversion to eventually become the world's foremost antibiotic hunter, H. Boyd Woodruff (1917–2017), one of his graduate students from that era recalled: "In 1939, four years before the discovery of streptomycin, I arrived at Dr. Waksman's office in the Rutgers Agriculture School as a new Ph.D. student. About a month after my arrival, Dr. Waksman excitedly came rushing into my laboratory and cried out 'Woodruff, Woodruff, drop everything. My former student René Dubos has discovered a way to find antibacterial agents produced by soil microorganisms. And he found an antibiotic; I am impressed. We must discover a better one' " (8).

The first antibiotic discovered by the Rutgers group was from one of Waksman's favorite soil microbes, the actinomyces, and then named it actinomycin. It was a potent antibacterial agent, but quite toxic to animals as well. It became however, the first antibiotic to be useful in cancer treatment. This success, stimulated Waksman to redirect most of his efforts to antibiotic hunting (9). Probably the most important of their discoveries came in 1943 (10). Streptomycin, isolated by Albert Schatz (1920–2005) from the microbe *Streptomyces griseus*, has been universally hailed as a milestone in antibiotic development. This antibiotic, which was rapidly developed, successfully treated both the Gram-positive and Gram-negative groups of bacteria, and importantly, was the first medication to successfully treat tuberculosis, the scourge of the ages (11). Anecdotally, it is of interest to note that the third patient treated with streptomycin in the U.S. and the first to be successfully treated (i.e., survived) was Robert Dole (1923–2021), a wounded veteran, who experienced a full recovery and later would go on to be Majority Leader of the U.S. Senate and candidate for President of the United States (12).

PENICILLIN

The story of the discovery of penicillin in 1929, the first "wonder drug," is legendary. Alexander Fleming finding a mold colony inhibiting lethal germs on a petri plate culture accidentally left uncovered on a window still is an apocryphal story told to the young to illustrate the chance nature of discovery coupled with insightful observation of such "accidents." Reality was not quite so dramatic. Fleming was studying various strains of the common bacteria of skin infections, *Staphylococcus*, and he was examining their long-term growth on culture plates. In his own words: "While working with staphylococcus variants a number of culture-plates were set aside on the laboratory bench and examined from time to time. In the examinations, these plates were necessarily exposed to the air [when the covers were removed to examine them] and they became contaminated with various micro-organisms. It was noticed that around a large colony of a contaminating mould the staphylococcus colonies became transparent and were obviously undergoing lysis" (13). In this very first publication in which Fleming studied the killing and growth-inhibiting effects of the substance emanating from this mold which he dubbed "penicillin," he also predicted its future as an anti-infective medicine: "It is suggested that it may be an effective antiseptic for application to, or injection into, areas infected with penicillin-sensitive microbes."

Alexander Fleming was a physician interested in infectious diseases, working in a pathology laboratory at St. Mary's Hospital in London where he was a professor of bacteriology at the University of London. He was well-aware of the fact that others had found various examples of microbial antagonisms and substances that could lyse bacteria (species-specific lytic substances called bacteriocines as well as the recently discovered viruses of bacteria called bacteriophages). Indeed, Fleming himself had previously discovered the lytic enzyme, lysozyme, a substance that dissolves many kinds of bacteria by breaking down their cell walls. Lysozymes of various sorts have been found in such diverse sources as tears and egg whites.

Penicillin seemed different, however. First, it was effective at incredible dilutions, and second, it was, even in high concentrations, non-toxic to animals. Sadly for Fleming, the technologies for finding high-producing strains of the mold and for purifying and concentrating the penicillin were in their infancies. And in addition, he was no biochemist, familiar with the properties of fragile biological chemicals. His attempts to obtain useful amounts of penicillin were rather dismal failures.

The second world war would provide the impetus to re-examine penicillin's potential. The need for some sort of treatment for the microbes such as staphylococci that infected war wounds was imperative. Early technologically oriented organic chemists and biochemists set out to make penicillin useful for scientific study. Ernst Chain (1906–1979), a young Cambridge-educated biochemist joined fellow bio-chemist, Howard Florey (1898–1968), a professor of pathology at Oxford, to study antibacterial substances produced by microbes. They were intrigued by Fleming's work on penicillin, and in 1940 they devised methods to purify and concentrate active penicillin, enough to start serious experimentation. While it is gratifying in retrospect to imagine altruistic, humanistic motivation to this work, Florey was more circumspect and in a reflective interview in 1967 he belied his own views as a basic scientist, seeking to discover nature's secrets: "People sometimes think that I and the others worked on penicillin because we were interested in suffering human-ity. I don't think it ever crossed our minds about suffering humanity. This was an interesting scientific exercise, and because it was of some use in medicine is very gratifying, but this was not the reason that we started working on it" (14).

In August of 1942 their team at Oxford provided Fleming with some of their first samples to treat a patient who was dying of streptococcal meningitis. Although the patient's infection had become resistant to all the sulfa drugs that were tried and the patient was moribund, the treatment with penicillin (one initial dose and then 6 days later daily doses for 9 days) showed immediate improve-ment and a month after the start of penicillin, the patient was discharged and went home cured (15). Clearly an amazing result, even recognized by an editorial in *The Times* (London) (16). Almost at the same time, the transatlantic scientific connec-tions between the microbiologists in the United Kingdom and those in the United States provided another early "miracle cure." As so often happens, personal con-nections belie formal ones. In March 1942 at Yale-New Haven Hospital in Con-necticut a 33 year-old woman was dying of blood poisoning by a streptococcal infection. Her doctor, John Bumstead (1897–1957), had another patient who was a colleague as well, Professor John Fulton (1899–1960), a famous neurophysiolo-gist. Fulton had been close friends with Howard Florey since their student days at Oxford, and indeed, Florey's children were living with the Fulton family in New Haven, safe from the wartime bombing in the U.K. Fulton knew of Florey's recent results with penicillin and Bumstead persuaded him to use his good offices to obtain some of the new drug for a desperate clinical trial on his moribund patient. Fulton managed, with his "connections" to obtain a small sample of the new drug which arrived in New Haven on 12 March 1942. The (then) young house doctor, Charles M. Grossman (1914–2014), who administered this penicillin, has published a detailed account of that historic event:

"The first vial was mailed to Dr. Bumstead in New Haven from Merck & Co. in Rahway, New Jersey, and I took the vial to Dr. Morris Tager, Associate Professor of Bacteriology and Immunology. We discussed what to do with the pungent, brown-red powder. 'We decided to dissolve it in saline and pass it through an E.K. Seitz [asbestos] filter pad to sterilize it,' wrote Dr. Tager in 1976 [ref. in original]. We then returned to the patient, and I injected 5000 U into the intravenous tubing. Rocko Fasanella, a medical student (later a professor of ophthalmology at Yale), gave subsequent doses every 4 hours. By Monday morning, the patient was eating hearty meals, the intravenous treatment was stopped, and she subsequently received 5000 U intravenously every 4 hours. On Monday morning rounds, Dr. Wilder Tileston, a senior consultant, looking at the temperature chart, muttered to those of us close enough to hear, "black magic." My fellow intern, Dr. Herbert Tabor (later the editor of the *Journal of Biological Chemistry* for many years), saved all of the patient's urine because Dr. Heatley had informed us that more penicillin could be purified from it than could be produced by cultivation.

Probably up to 95% of each intravenous dose was excreted unchanged. When Dr. Heatley delivered a subsequent vial, some of which had come from the patient's urine, he carted the gallons of urine back to Rahway. The patient survived and later died of other causes at the age of 90 years" (17).

"Black magic" indeed.

Early studies on the use of penicillin showed that while injected penicillin found its way into the blood stream and thus to the sites of infection, it was also rapidly eliminated by the kidney in an unchanged form. This was both a curse and blessing. The very rare and precious drug had to be repeatedly injected at frequent intervals and at relatively high doses, but the excreted drug could also be recovered from the patient's own urine for recovery and reuse. As Fleming noted later: "to keep a reasonable concentration in the blood, injections have to be repeated every few hours. Florey has compared it to filling a bath by turning on the water-taps and leaving the plug out" (18).

Penicillin, however, was in extremely short supply and the Oxford biochemists were unable to find ways to scale up production in the quantities needed for widespread testing, let alone war time clinical application. In the early days of WWII, the priorities of the U.K. government did not include research on penicillin, but arrangements with the U.S., with experience in fermentation technologies, provided a possible way forward. The story of the Anglo-American collaboration to develop penicillin is well-known. Florey and a colleague arrived in the U.S. in 1941 and headed to the U.S. Department of Agriculture Northern Regional Research Laboratory in Peoria, Illinois, (far from the vulnerable coasts of America) where they enlisted U.S. microbiologists to set up large-scale fermentation facilities and get to work on mass-producing penicillin.

The mycologists (mold specialists) at Peoria realized that the Oxford strains of *Penicillium* did not produce very much penicillin even under the optimal conditions discovered by Florey and Chain. They set out to find higher yielding strains of the mold. These mycologists examined molds from "a wide variety of sources including (i) moldy food products such as bread, cheese, cured meats, home-canned fruits and vegetables, etc.; (ii) fresh fruits and vegetables in the earlier stages of spoilage; and (iii) fertile cultivated soils collected from various stations in the United States and from foreign countries, including Newfoundland, England, Mexico, Panama, Cuba, Brazil, Australia, and India" (19). As luck would have it, the prize strain which produced a fifty-fold increase in penicillin yield was found on an over-ripe cantaloupe from a Peoria fruit market near the laboratory. The exact account of this discovery is lost to history, but one legend suggests that one lab member had been charged with daily searches of fruit and vegetable markets looking for moldy produce. "Moldy Mary" (so nicknamed by the press) was actually Mary K. Hunt, a member of the research team, who was formally acknowledged in the publication of this work: "We are likewise indebted to Miss Mary K. Hunt for collecting samples of moldy materials and for assisting in the isolation and preliminary testing of many strains" (20). Legend also has it that after cutting off the mold for culture, the remainder of the melon was devoured by the lab staff members.

The collaboration between the National Research Council in the United Kingdom and the Northern Research Lab of the USDA in the United States resulted in rapid progress in bringing penicillin into widespread use. Not only were higher producing strains of *Penicillium* introduced, but improved methods of large-scale fermentations were devised. All in all, yields were rapidly increased nearly a hundred-fold. Still, penicillin continued to leak out of patients by urinary excretion at an alarming rate. Very soon, this problem would be addressed in two ways: known drugs were employed to block kidney excretion leading to prolonged blood levels, and additionally, chemical modifications of purified penicillin were discovered that slowed kidney excretion without affecting its antibacterial actions.

Penicillin was indeed a "wonder drug." It attacked a very broad range of troublesome microbes, those termed "Gram positive," a classification based on the property that the cell walls of this group of organisms would turn blue with a chemical stain called crystal violet, a method introduced in 1884 by the Danish scientist Christian

Gram (1853–1938). The "Gram positive" bacteria include those that frequently cause wound infections (*Staphylococcus*), pneumonia (*Streptococcus pneumoniae*), diphtheria, anthrax, urinary tract infections (*Enterococcus*), and a host of other troublesome bacteria. Penicillin was found to treat some other Gram-positive infections such as the ancient scourges of plague and syphilis as well. By the mid-1940s, it was being widely hailed as "a potent new chemotherapeutic agent" (21).

THE GOLDEN AGE

It is indeed appropriate to view the era just after the introduction of streptomycin and penicillin as the golden age of antibiotics: the next new antibiotic to be discovered was a golden yellow substance isolated, again, from a species of Waksman's favorite microbe, actinomyces. Aureomycin (L. *aureus*: gold) was isolated from *Streptomyces aureofaciens* in the commercial firm, Lederle Laboratories, by a team headed by Yellapragada Subbarow (Subba Rao) (1895–1948) and Benjamin Duggar (1872–1956) (22). This antibiotic was one of the first "broad spectrum" antibiotics; it was effective against many different bacterial classes, indeed a "magic bullet" that hit almost everything at which it was aimed. Chemically determined to be a molecule with four rings of carbon atoms, it defined a new kind of antibiotic called tetracycline. Aureomycin, a trade name, was, in pharmacological terms, chlortetracycline.

When it was introduced in 1945, Aureomycin became the next "go-to" wonder drug for infectious diseases. Since it was "broad spectrum" it was less important to identify the infecting organism, just prescribe Aureomycin, and the cure would follow. It did not show some of the troubling side effects of streptomycin and was active on many organisms that penicillin did not touch. Millions of patients were treated with this new wonder drug. Even to this day, some of those early patients still carry the evidence of this era because it was only later discovered that the golden-drug was incorporated into the growing teeth of children providing permanent yellow staining of their teeth. With appropriate dosing and judicious use, tetracyclines, while one of the earliest antibiotics, still remain an important class of antibiotics.

Another example from the early days of the antibiotic gold rush is a substance isolated from another actinomyces strain, *Streptomyces venezuelae*, in 1947 by Paul Burkholder (1903–1972) at Yale and his associates at the Parke-Davis pharmaceutical company (23). Under the trade name Chloromycetin with the chemical name chloramphenicol, this drug was found to be another broad-spectrum antibiotic. Its molecular structure, however, was much simpler than that of either penicillin or the tetracyclines and in 1949 chloramphenicol became the first antibiotic to be commercially produced by chemical synthesis rather than as a byproduct of microbial fermentations (24). While initially chloramphenicol was widely used for many infections, its major side effect, suppression of blood formation in the bone marrow, has limited its current use to cases where it is the last resort, or where it is applied only topically.

These successes of the early antibiotic hunters spurred on the work of the major pharmaceutical companies in the search for new antibiotics with more potent activities against both common and exotic infections and with fewer side effects. Several new classes of antibiotics were discovered and introduced to the market in the decades of the 1950s and 1960s. In the three decades from 1940–1970 there were eleven new classes of antibiotics introduced, but in the three decades from 1970–2000 only two new classes were discovered (25).

By 1970 we had a collection of antibiotics to treat many of the traditional infectious diseases, mostly bacterial diseases. Chemotherapy for viruses was almost non-existent and for fungus diseases woefully few. Still, the "wonder drugs" (and advances in sanitation and public health) were being hailed as our "conquest" of acute infectious diseases. What only a few observant scientists noted, however, was the problems we were creating by this apparent conquest. Their concerns were two-fold, both relating to the widespread presence of antibiotics in the environment: first was the medical overuse of antibiotics in inappropriate circumstances; second was the massive use of antibiotics to suppress microbial populations in agriculture to increase yields of animal products. Basically, too much of a good thing.

Mid-twentieth century biochemistry was flush with success in the field of nutrition: many essential nutrients were being identified, the importance of vitamins had made its way into the public discourse, and human and animal diets were being optimized for growth and health. Agricultural scientists studied the way soils, animal feeds, genetic breeding, and cultivation practices could increase crop and animal yields. The role of microbes in animal and plant health was also of interest. So-called "germ-free" animals were produced in laboratories to allow experimenters to better understand the newly found relationship between an animal and the microbes it might harbor. It was discovered, for example, that certain intestinal microbes produced substances that the animal (and even the human animal) needed for life but could not produce itself. One study to examine these relationships provided evidence that adding antibiotics to the diet would increase the growth and production of body mass in farm animals. Thus was born the great global antibiotic dispersion that would deliver better yields but also dire consequences for human health.

In the late 1940s, scientists at Lederle Laboratories found a substance produced by microbial fermentation, when fed to chicks, would significantly promote weight gain and growth. "The term 'animal protein factor' has been used for the dietary factor which is needed for the growth of chicks on diets consisting principally of corn and soy bean meal" (26). For reasons that are obscure, the source of the "animal protein factor" was deep aerobic cultures of the organism *Streptomyces aureofaciens*. And it worked for piglets, too. As it turned out, "animal protein factor" was none other than Aureomycin (27). Since this new antibiotic was becoming available in industrial amounts, it was tested in a variety of animals and soon became a common supplement in poultry feed. By 1956 the exploration of the use of antibiotics in animal nutrition had become widespread and the National Research Council in the U.S. convened a conference to assess the state of the research. Four potential mechanisms for the growth-promoting effects of antibiotics (mostly Aureomycin) were proposed by Carl A. Baumann (1896–1969), a biochemist from the University of Wisconsin:

"(a) the antibiotic suppresses organisms that cause "disease," usually too mild to be recognized as such but sufficient to depress the growth rate below that which would occur in the absence of the disease;

(b) the antibiotic changes the microorganisms in such a way that vitamin synthesis is increased in those parts of the digestive tract in which microbial vitamin can be made available to the host;

(c) the antibiotic suppresses organisms that normally destroy vitamins or other nutrients in the intestine, or that absorb them and thus render them unavailable to the host;

(d) the presence of antibiotics tends to make the intestines thinner than those of control animals. [and]...thinner intestines might offer one means by which improved absorption of nutrients or better net usage of nutrients is brought about" (28).

A decade later in 1966, a similar conference was less focused on the positive effects of antibiotics in agriculture and raised a persistent concern about the effects of antibiotics in foodstuffs on human health and the appearance of antibiotic resistance in both animal and human microbes (29). The specter of things to come was becoming visible just over the scientific horizon.

NOTES AND REFERENCES

1. **Domagk G.** 1935. Ein Beitrag zur Chemotherapie der bakterillen Infektionen. *Dtsch Med Wochenschr* **61**:250–253.
2. The "Winogradsky column" is a simple device for culturing a large diversity of microorganisms. It is a column of pond mud and water mixed with a mixture of materials providing a carbon source, a sulfur source and other organic materials. Incubating the column in sunlight for months results in an aerobic/anaerobic gradient as well as a sulfide gradient. These gradients promote the growth of different microorganisms such as *Clostridium, Desulfovibrio, Chlorobium, Chromatium, Rhodomicrobium,* and *Beggiatoa* in different regions of the column based on their adaptation to different ecological niches. Depending on added nutrients, a variety of organisms can be demonstrated.

3. "Think globally, act locally" has been credited to Dubos although this attribution is disputed. See Willy Gianinazzi W. 2018. Penser global, agir local. Histoire d'une idée. *Rev crit d'écolog polit* **46**:24, who quotes **Dubos R.** 1977. The despairing optimist. *Am Schol* **Spring**:156.

4. **Dubos R, Avery OT.** 1931. Decomposition of the capsular polysaccharide of pneumococcus type iii by a bacterial enzyme. *J Exp Med* **54**:51–71.

5. **Dubos RJ.** 1939. Studies on a bactericidal agent extracted from a soil bacillus: I. Preparation of the agent. Its activity in vitro. *J Exp Med* **70**:1–10.

6. **Waksman SA, Starkey RL.** 1923. Partial sterilization of soil, microbiological activities and soil fertility. *Soil Sci* **16**:343–358.

7. **Waksman SA.** 1937. Associative and antagonistic effects of microorganisms. I. Historical review of antagonistic relationships. *Soil Sci* **43**:51–68.

8. **Woodruff HB.** 2014. Selman A. Waksman, winner of the 1952 Nobel Prize for physiology or medicine. *Appl Environ Microbiol* **80**:2–8.

9. **Waksman SA, Horning E.** 1943. Distribution of antagonistic fungi in nature and their antibiotic activity. *Mycologia* **35**:47–65.

10. **Schatz A, Bugie E, Waksman SA.** 1944. Streptomycin, a substance exhibiting antibiotic activity against Gram-positive and Gram-negative bacteria. *Exp Biol Med (Maywood)* **55**:66–69.

11. **Marshall G, Blacklock JWS, Cameron C, Capon NB, Cruickshank R, Gaddum JH, Heaf FRG, Hill AB, Houghton LE, Hoyle JC, Raistrick H.** 1948. Streptomycin treatment of pulmonary tuberculosis: A Medical Research Council investigation. *BMJ* **2**:769–782.

12. **Cramer RB.** 1992. *What it Takes: The Way to the White House* p 110–111. Random House, New York, New York.

13. **Fleming A.** 1929. On the antibacterial action of cultures of a penicillium, with special reference to their use in the isolation of *B. influenzae. Br J Exp Pathol* **10**:226–236.

14. de **Berg H.** 1967. *Transcript of Taped Interview with Lord Howard Florey,* 5 April 1967, National Library of Australia, Canberra, p. 9 of 15. (cited in Denise Sutherland and Elissa Tenkate, (19 February 1998). "Howard Walter Florey". Australian Nobel Laureates. University of Melbourne. Retrieved 3 Oct. 2022. https://www.asap.unimelb.edu.au/bsparcs/exhib/nobel/florey.htm

15. **Fleming A.** 1943. Streptococcal meningitis treated with penicillin: measurement of bacteriostatic power of blood and cerebrospinal fluid. *Lancet* **242**:434–438.

16. **Anonymous.** 1942. Penicillium. *Times Lond* **27**:5.

17. **Grossman CM.** 2008. The first use of penicillin in the United States. *Ann Intern Med* **149**:135–136.

18. **Fleming A.** 1944. Penicillin: The Robert Campbell Oration. *Ulster Med J* **13**:95–108, 2.

19. **Raper KB, Alexander DF, Coghill RD.** 1944. Penicillin: II. Natural variation and penicillin production in *penicillium notatum* and allied species. *J Bacteriol* **48**:639–659.

20. See reference 19.

21. **Myers WG.** 1944. Review papers: Penicillin: A potent new chemotherapeutic agent. *Ohio J Sci* **44**:278–286.

22. **Harned BK, Cunningham RW, Clark MC, Cosgrove R, Hine CH, McCauley WJ, Stokey E, Vessey RE, Yuda NN, Subbarow Y.** 1948. The pharmacology of duomycin. *Ann N Y Acad Sci* **51**(Art. 2): 182–210.

23. **Ehrlich J, Bartz QR, Smith RM, Joslyn DA, Burkholder PR.** 1947. Chloromycetin, a new antibiotic from a soil actinomycete. *Science* **106**:417.

24. **Controulis J, Rebstock MC, Crooks HM.** 1949. Chloramphenicol (Chloromycetin). V. Synthesis. *J Am Chem Soc* **71**:2463–2468 http://dx.doi.org/10.1021/ja01175a066. The sulfa drugs were also synthetic, but are not considered "antibiotics" by many authors since they are not the product of biological processes.

25. **Conly J, Johnston B.** 2005. Where are all the new antibiotics? The new antibiotic paradox. *Can J Infect Dis Med Microbiol* **16**:159–160.

26. **Stokstad ELR, Jukes TH, Pierce J, Page AC** Jr, **Franklin AL.** 1949. The multiple nature of the animal protein factor. *J Biol Chem* **180**:647–654.

27. See reference 26; also **Cunha TJ, Burnside JE, Buschman DM, Glasscock RS, Pearson AM, Shealy AL.** 1949. Effect of vitamin B12, animal protein factor and soil for pig growth. *Arch Bioch* **23**:324–326; and **Stokstad E, Jukes TH.** 1959. Further observations on the 'animal protein factor.' *Proc Soc Exp Biol Med* **73**:523–528.

28. **Baumann CA.** 1956. Test Animals, p 47–54. In National Research Council, *Proceedings, First International Conference on the Use of Antibiotics in Agriculture.* The National Academies Press, Washington, DC.

29. **Somogyi JC, Francois AC (eds).** 1968. *Antibiotics in Agriculture. Proceedings of the Fifth Symposium of the Group of European Nutritionists in Jouy-en-Josas, April 1966.* S. Karger, Basel, Switzerland.

4 Hints of Trouble

With the success of microbe hunting in the first half of the twentieth century and the discovery of highly effective antimicrobial drugs in the 1940s and 1950s, microbiologists were feeling confident that the threats of infectious diseases, both to humankind and other animals would soon be a thing of the past. A famous infectious disease expert reassured us in 1963 that "Eradication has been demonstrated many times to be entirely practical within certain limits, even with the techniques of today. [...] Therefore we can look forward with confidence to a considerable degree of freedom from infectious diseases at a time not too far in the future. [...] then it seems reasonable to anticipate that with some measurable time, such as 100 years, all major infections will have disappeared" (1). But there were some scientists, perhaps thought to be Cassandras, who foresaw a more problematic future. One was René Dubos, one of the early discoverers of antibiotics. By 1965 he was looking back on the innocence of the 1950s...

"By the 1950's the most optimistic dreams of the founders of medical microbiology had been essentially fulfilled in several countries of Western civilization. A very large percentage of microbial agents of disease had been isolated, identified and cultivated in artificial media or in tissue cultures; bacteriologic purity of the food and water supplies had become possible through technologic advances; practical procedures had been worked out for the large-scale production of killed or attenuated vaccines; highly effective drugs had become available for the treatment of bacterial and parasitic infections; a variety of pesticides had been synthesized and had proved their usefulness for the control of insect vectors. [...] Therefore, most clinicians,

Magic Bullets, Miracle Drugs, and Microbiologists: A History of the Microbiome and Metagenomics,
First Edition. William C. Summers.

and public health officers, epidemiologists and microbiologists felt justified in proclaiming during the 1950s that the conquest of infectious diseases had finally been achieved. [...] The more important reason [rather than the emergence of resistances] for the stubborn persistence of infection lies in the obscure phenomenon concerning the relationships between man and his biologic environment" (2).

This prescient warning was about to come into full elaboration in the two decades that followed. The rise of drug resistance was some of the first evidence of our hubris concerning our understanding of microbial diversity and the crucial interactions in complex microbial communities that would be central to later microbiology in the twenty-first century.

MICROBIAL DRUG RESISTANCE

Just a decade after the introduction of the first antibiotics, and their burgeoning use in agriculture as well as medicine, the prescient René Dubos, delivering the dinner talk at the first international conference on antibiotics in agriculture in 1956, sounded a warning:

"Among scientific subjects, few are better suited to table conservation and after dinner speeches than the field of antibiotics. The method of discovery of these drugs involves no abstract concept; their application to the treatment of disease has strong emotional appeal; *the illusion that they will bring about the final conquest of infection is still widespread.* Finally, the introduction of antibiotics in medicine has a picturesque history made up of quaint reports of ancient folk remedies, and of romantic episodes in the life of modern scientists. Indeed, the whole subject evokes the mysterious atmosphere of alchemy or even magic, rather than the cold discipline of experimental science. *Note how often the word "miracle" is used in popular writings on antibiotics, and the tendency to regard as magicians those concerned with their discovery and use*" [italics added] (3).

Almost as soon as it was known that microorganisms could be killed by certain substances, it was recognized that some microbes could survive normally lethal doses and were described as "drug-fast" (German: *-fest* = -proof, as in *feuerfest* = fireproof; hence "drug-proof," in common usage by at least 1913). These early studies conceived of microbial resistance in terms of "adaptation" to the toxic agents (4). By 1907, Ehrlich more clearly focused on the concept of resistant organisms in his discussion of the development of resistance of the organism that causes African sleeping sickness (*Trypanosoma brucei*) to the lethal dye *p*-roseaniline (5). In 1911 it was reported that the germ of pneumonia could develop resistance to optochin, a quinine derivative (6). For every new agent that killed or inhibited microorganisms, resistance became an interest as well.

While we think of antibiotic resistance as a phenomenon of recent concern, the basic conceptions of the problems, the controversies, and even the fundamental mechanisms were well-developed in the early decades of the twentieth century. These principles were, of course, elaborated in terms of resistance to microbial toxins, such as the arsenicals, dyes, such as trypan red, and disinfectants, such as acid, phenols, and the like. However, by the time the first antibiotics were employed in the 1940s and resistance was first observed, the framework for understanding this phenomenon was already in place.

DRUG-FASTNESS

Drug-fastness became a topic of importance as microbiologists sought understanding of the growth, metabolism, and pathogenicity of bacteria, protozoa, and fungi. In 1913, Paul Ehrlich clearly described the basic mechanisms of drug action on microbes: "parasites are only killed by those materials to which they have a certain relationship, by means of which they are fixed by them" (7). He went on to describe specific drug binding (fixation) to specific organisms and elaborated "The principle of fixation in chemotherapy."

Once this principle was accepted, one could investigate how drugs are bound by microbes, what kinds of cross-sensitivities existed, and what happened when organisms became resistant to chemotherapeutic agents. Ehrlich noted that both trypanosomes (sleeping sickness) and spirochaetes (syphilis), his favorite experimental organisms, exhibited different chemoreceptors that were specific for drugs of a given chemical class. Thus, there seemed to be a chemoreceptor for arsenic compounds (arsenious acid, arsanilic acid, and arsenophenylglycine) that differed from the receptor for azo-dyes (trypan red and trypan blue) as well as from the receptor for certain basic triphenylmethane dyes, such as fuchsin and methyl violet.

Resistance, therefore, was readily explained as "a reduction of their [the chemoreceptors] affinity for certain chemical groupings connected with the remedy [the drug], which can only be regarded as purely chemical" (8). Clearly, Ehrlich's approach was an outgrowth of his earlier work on histological staining and dye chemistry and reflected his strong chemical thinking.

Already in 1913, the problem of clinical drug resistance was confronting the physician and microbiologist. Ehrlich discussed the problem of "relapsing crops" of parasites as a result of the parasites' biological properties. His views were mildly selectionist, but he also held the common view that microbes had great adaptive power and that the few, which managed to escape destruction by drugs (or immune serum), could subsequently change into new varieties that were drug-fast or serum-proof.

One corollary of the specific chemoreceptor hypothesis was that combined chemotherapy was best carried out with agents that attack entirely different chemoreceptors of the microbes. Ehrlich, who frequently resorted to military metaphors, wrote: "It is clear that in this manner a simultaneous and varied attack is directed at the parasites, in accordance with the military maxim: 'March apart but fight combined' " (9). He also allowed for the possibility of drug synergism so that in favorable cases the effects of the drugs may be multiplied rather than simply additive. From the earliest days of chemotherapy, it appears that multiple drug therapy with agents with different mechanisms was seen as a way to circumvent the problem of "relapsing crops" of resistant organisms.

Ehrlich, too, realized the relationship between evolution of resistant variants and the dose of the agent used to treat the infection. Clinical practice often used remedies in increasing dosages, perhaps a therapeutic principle derived from empirical treatment practice of long tradition. He noted that these were precisely the conditions likely to lead to emergence of drug-fast organisms and developed the idea of *"therapia sterilisans magna"* (total sterilization) in which he advocated the maximum microbicidal dose that was non-toxic to the host (10). Indeed, by 1916, there was experimental confirmation, in controlled *in vitro* laboratory studies, that gradual increases in drug concentration would lead to outgrowth of resistant spirochetes, while exposure to initial high concentrations of anti-syphilis agents (arsenicals, mercuric, and iodide compounds) would not (11).

As research on antibiotic sensitivities and resistance progressed, standard microbiological techniques were adapted to these new topics. One method, devised by Vernon Bryson and Wacław Szybalski, the gradient plate technique demonstrated both comparative resistance as well as genetic variants that are present in a supposedly homogeneous population of bacteria (12) (Figure 4.1).

DISINFECTION

Often early research on antimicrobial agents was directed to problems of "disinfection" and related matters of public health, and the origins and properties of resistant organisms became of concern in the "fight against germs" (13). Protocols for inducing drug-resistance *in vivo* were elaborated, and the relevance of laboratory culture tube resistance to "natural" resistance occurring in the infected animal was debated in the literature of the 1930s and 1940s.

MICROBIAL METABOLISM AND ADAPTATION

The basic issue, as we would see it today, that faced microbiologists in the early days of antimicrobial research is one of adaptation versus mutation. It was passionately debated and contested by leading microbiologists from the mid-1930s until the early 1960s. Even those who viewed most microbial resistance as some sort

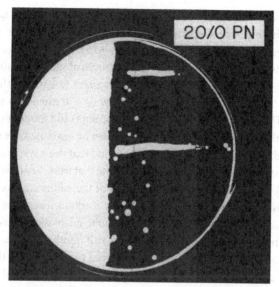

FIGURE 4.1 *Gradient plate showing bacteria evenly spread on medium in which the concentration of an antibiotic varied in a linear gradient from one side of the dish to the other (35). Most of the bacteria could grow at low concentrations, but only a few, now known to be mutant forms grew beyond the concentration that inhibited the bulk of the population.*

of heritable change, or mutation, were divided on the basic problem of whether the mutations arose in response to the agent, or occurred spontaneously and were simply observed after selection against the sensitive organisms. This problem was unresolved until the 1940s and 1950s.

As early as the 1920s, the ability of bacterial cells to undergo infrequent, abrupt, and permanent changes in characteristics was interpreted as due to the phenomenon of mutation as had been described in higher organisms (14). The relation of these mutations to the growth conditions where they could be observed, was, however, unclear. In the 1930s, this question was confronted directly by Isaac M. Lewis (1878–1943) who studied the mutation of a strain of "*Bacillus coli mutabile*" (*Escherichia coli*) from the inability to use the sugar, lactose, to lactose-utilizing proficiency (15). Lewis laboriously isolated colonies and found that even in the absence of growth in lactose, the ability to ferment this sugar arose spontaneously in about one cell in 100,000. This work was the beginning of a long line of investigations that quite conclusively showed that mutation is (almost always) independent of selection.

The second kind of adaptation, a kind "due to chemical environment," is of special historical interest. As early as 1900, Frédéric Dienert (1874–1948) found that yeast that were grown for some time in galactose-containing medium became adapted to this medium and would grow rapidly without a lag when sub-cultured into fresh galactose medium, but that this "adaptation" was lost after a period of growth in glucose-containing medium (16). By 1930, Hennig Karström (1899–1989) in Helsinki had found several instances of such adaptation (17). For example, he found that a strain of *Klebsiella aerogenes* (a bacteria that can cause opportunistic infections in humans) could grow on ("ferment" to use the older term) xylose if "adapted" to do so, but that this strain could ferment glucose "constitutively" without the need for adaptation. When he examined the enzyme content of these adapted and unadapted cells, he found that there were some enzymes that were "constitutive" (always present) and some that were "adaptive" (only there when needed). Thus, the metabolic properties of the culture mirrored the intracellular chemistry. By experiments in which the medium was changed in various ways, Karström and others showed that metabolic adaptation could sometimes take place even without measurable increase in cell numbers in the culture. Marjory Stephenson (1885–1948), a leading bacterial physiologist in the middle of the twentieth century, described these variations in her influential book, *Bacterial Metabolism*, as "Adaptation by Natural Selection" and "Adaptation due to Chemical Environment" (18). The former included the phenomenon that is now termed mutation.

Between 1931 and the start of WWII, Stephenson and her students, John Yudkin (1910–1995) and Ernest Gale (1914–2005), investigated bacterial metabolic variation in detail, often exploiting the lactose-fermenting system in intestinal bacteria to study it. The mechanism of chemical adaptation, however, eluded them. The final paragraph of her monograph expressed her belief in the importance of the study of bacterial metabolism: "It (the bacterial cell) is immensely tolerant of experimental meddling and offers material for the study of processes of growth, variation and development of enzymes without parallel in any other biological material" (19).

In 1934, another research group on "bacterial chemistry" consisting of Paul Fildes (1882–1971) and B.C.J.G. Knight (1904–1981) was established at Middlesex Hospital in London (20). Fildes and Knight investigated bacterial nutrition and established vitamin B1 (thiamine) as a growth factor for *Staphylococcus aureus* (the notorious "golden staph" of many human infections). One recurrent theme in their early work was the finding that they could "train" bacteria to grow on media deficient in some essential metabolite. For example, they could train *Bact. typhosum* (modern name: *Salmonella enterica* serovar Typhi) to grow on medium without tryptophan or without indole. Fildes noted that "during this time little attention was given to the mechanism of the training process, but it was certainly supposed that the enzyme make-up of the bacteria became altered as a result of a stimulus produced by the deficiency of the metabolite" (21).

By the mid-1940s, however, Fildes and his colleagues undertook a study of the mechanism of this ubiquitous "training." Was it another example of enzyme adaptation or was it something else? Using only simple growth curves, viable colony counts on agar plates, and ingenious experimental designs, they concluded, "that 'training' bacteria to dispense with certain nutritive substances normally essential may be looked upon as a cumbersome method for selecting genetic mutants" (22). Little by little, the underlying mechanisms of the different kinds of biochemical variations seen in bacteria were becoming clear, and little by little, genetics was joining biochemistry as a powerful approach to study bacterial physiology. This understanding, of course, was central to discovering the underlying mechanisms involved in the variation of microbial behavior related to drug resistance.

ADAPTATION OR MUTATION?

With the discovery of antibiotics and their medical applications, drug resistance took on new relevance and new approaches became possible. No sooner were new antibiotics announced than reports of drug resistance appeared: sulfonamide resistance in 1939; penicillin resistance in 1941 and streptomycin resistance in 1946, to cite a few early reports in the widely-read literature (23). Research on resistance focused on three major problems: (i) cross-resistance to other agents, that is, was resistance to one agent accompanied by resistance to another agent? (ii) distribution of resistance in nature, that is, what was the prevalence of resistance in naturally occurring strains of the same organism from different sources? (iii) induction of resistance, that is, what regimens of drug exposure led to the induction or selection of resistant organisms?

While many practically useful results came from such research, two lines of investigation emerged that were to later prove scientifically interesting. Drug resistance provided a potent experimental tool to microbiologists who were studying bacterial genes and mutations, since it allowed the analysis of events that took place at extremely low frequencies. In 1952 Joshua Lederberg (1925–2008) and Esther Lederberg (1922–2006) were able to use both streptomycin resistance and their newly-devised replica plating technique to provide direct and convincing evidence to support the belief that mutations to drug resistance occurred even in the absence of the selective agent (24). Not only did such work on drug resistance clarify the nature of microbe-drug interactions, but it provided a much-needed tool to the nascent field of microbial genetics (25). Just as Paul Ehrlich's 1913 summary of the principles of chemotherapy provided a window on early understanding of drug resistance, we can find a similar succinct presentation of the mid-twentieth century state of the field in a review by Bernard Davis (1916–1994) in 1952 (26). By this time, genetics of microbes had replaced microbial biochemistry as the fashionable mode of explanation for bacterial drug resistance as well as an increasingly useful tool for microbial identification and classification. Although bacteria do not have a cytologically visible nucleus with stainable chromosomes, it was recognized that they have "nucleoid bodies" and that the material in this structure appeared to behave in a way similar to the chromosomes of higher organisms. By mid-century, bacteria had become "real" cells, with conventional genetic properties. If bacteria were like higher organisms, and since "almost all the inherited properties of animals or plants are transmitted by their genes" it was only logical, Davis argued, to consider genetic mutations as the basis for inherited drug resistance. He concluded that the recent work in microbial genetics by Salvador Luria (1912–1991) and Max Delbrück (1906–1981) (27), by Lederberg and Lederberg (28), and by Howard B. Newcombe (1914–2005) (29), settled the matter: the mutations to drug resistance were already present, having originated by some "spontaneous" process, and were simply selected by the application of the drug. Needless to say, these results were widely hailed by Darwinists as laboratory proof that organic evolution as envisioned by Charles Darwin was no longer "just a theory," it was real.

A very important clinical correlate of this new understanding of the nature of bacterial drug resistance was its application to combination chemotherapy. Since it became clear that mutations to resistance to different agents were independent events, the concept of multiple drug therapy, initially envisioned by Ehrlich, was refined and made precise. It was realized that adequate dosages and lengths of treatment were necessary if the emergence of resistant organisms was to be avoided (30).

MULTIPLE DRUG RESISTANCE AND CROSS RESISTANCE

In the 1950s, in the era of many new antibiotics and the emphasis on surveys of both cross resistance and distributions of resistance in natural microbial populations, especially in Japan, it was recognized that many strains with multiple drug resistances were emerging. The appearance of such multiple drug resistance could not be adequately explained on the basis of random, independent mutational events. Also, the patterns of resistance were complex and did not fit a simple mutational model. For example, resistance to chloramphenicol was rarely, if ever, observed alone, but it was common in multiply-resistant strains. Careful epidemiological and bacteriological studies of drug resistant strains in Japan led Akiba et al. and Ochiai et al. to suggest that multiple drug resistance may be transmissible both *in vivo* and *in vitro* between bacterial strains by so-called resistance transfer factors (RTFs or R-factors) (31).

Genetic analysis of this phenomenon showed that the genes for these antibiotic resistance properties resided on the bacterial genome, yet were transmissible between strains albeit at low frequency. Further study showed that the transfer of these genes was mediated by a bacterium-to-bacterium movement of a segment of DNA called a plasmid (a process dubbed conjugation or horizontal gene transfer; more about this later) (32). Because of the promiscuous nature of RTFs, once a gene for drug resistance evolves in one microbe, it can rapidly spread to other organisms. Interestingly, because the R-factor plasmids replicate to high copy number, probably as a way to provide high levels of the drug-resistant protein, these plasmids have become the molecule of choice for molecular cloning and manufacturing biotechnology.

With the better understanding of the genetics of drug resistance and the classification of the types of resistance, the biochemical bases for resistance were elucidated. Knowledge of the mechanism of action of an agent led to understanding of possible mechanisms of resistance. The specific role of penicillin in blocking cell wall biosynthesis, coupled with the knowledge of the structures of bacterial cell walls, could explain the sensitivity of Gram-positive organisms and the resistance of Gram-negative organisms to this antibiotic. Likewise, understanding of its metabolic fate led to the finding that penicillin was often inactivated by degradation by an enzyme called beta-lactamase, which provides one mechanism of bacterial drug resistance. Detailed biochemical studies of the actions of antimicrobials have led to the understanding of the many ways in which microbes evolve to become resistant to such agents.

ANTIBIOTIC RESISTANCE IN THE CURRENT CONTEXT

The successes of wonder drugs against infectious diseases in the 1950s and 1960s exposed several unanticipated circumstances. These were scientific, environmental, and economic. First, the fact that many of the newly discovered antibiotics were "broad spectrum" led physicians to become casual about the need to identify the particular offending pathogenic microbe and prescribe the appropriate therapy. The practice of identifying a pathogen by laboratory cultures followed by checking its sensitivity to a proposed therapy gave way in everyday practice to the "therapeutic trial" approach: treat with a broad spectrum antibiotic and if the patient is cured, the identification of the pathogen was of only "academic interest." A corollary of this approach is that antibiotics became widely used for diseases that were not bacterial but viral in origin. The antibiotics in common use were known to be ineffective against viral diseases because viruses are of a totally different biological realm. This practice inadvertently increased the antibiotic load in the biological environment leading to more unnecessary selective pressure on the normal microbial populations. Even if antibiotic resistance developed in innocuous bacteria, the phenomenon of lateral gene transfer meant that the resistance gene soon spread like a microbial epidemic into the more troublesome germs that would soon become a clinical problem.

A phenomenon associated with casual antibiotic use had been foreseen by Ehrlich himself, the problem of insufficient dosages that then allowed for step-wise resistance to develop. Antibiotics are often so effective that a patient feels wellquite rapidly, assumes the infection is eradicated, and stops taking the medication. In reality, the patient's improvement is only the result of reduction, not elimination, in the infection, and the reduced population of remaining bacteria are enriched in variants that are relatively more resistant to the antibiotic. An ideal Darwinian selection of the fittest to develop resistant microbial strains.

Lateral gene transfer has also led to the movement of antibiotic resistance genes from animal microbes into human microbes. The widespread use of antibiotics in farming has been a major source of resistance to newly introduced antibiotics. Most antibiotics are produced by microbial fermentations in large liquid cultures on an industrial scale. The microbes often secrete their antibiotics into the liquid broth from which the antibiotic is then chemically separated. The left-over residue of the microbes themselves as well as the depleted broth are frequently processed into animal feed. This residue is rich in essential nutrients such as vitamins, amino acids, and other compounds that promote growth in farm animals. Not surprisingly, this residue also contains some level of the antibiotic that is being produced that then finds its way into the animal being fed. Again, an ideal setup for inadvertent selection of bacterial resistance genes to flourish (33).

The last, but not least, context in our consideration of antibiotic resistance is an economic problem: how can financial incentives to hunt for new antibiotics be maintained in the face of persistent and rapid obsolescence? It is estimated that the costs of discovering and developing a new antibiotic for clinical use is of the order of a billion U.S. dollars (34). A new antibiotic, to recover this scale of investment, must be widely sold and used. Yet this very wide usage is certain to lead to development of microbial resistance which renders the drug impotent. A true dilemma. Use it and you destroy your market. Save it only for essential use and you go broke before you can recover your investment. This dilemma underlies one of the essential themes of this book: the advent of the antibiotic revolution brought with it the dubious optimism that "there will always be a new, 'next generation' antibiotic to allow us to outrun the microbe's ability to develop resistance to our current drugs."

In this chapter we have seen how the discovery and widespread use of antibiotics in medicine and agriculture demonstrated both the diversity and mutability of microbes. The usual and frequent emergence of antibiotic resistance showed us the downside of antibiotic treatments and from its beginning warned of the dangers of broad failures to come. The study of antibiotic resistance and sensitivity allowed for better methods to characterize microbes and isolate new species, as well as the development of tools for gene manipulation. Microbiological faith would soon be challenged and scientific hubris would be on display as new microbes emerged and the magic bullets of antibiotics lost their wonderous charm.

NOTES AND REFERENCES

1. **Cockburn A.** 1963. p 149–150. *In The Evolution and Eradication of Infectious Diseases.* Johns Hopkins University Press, Baltimore, Maryland.
2. **Dubos RJ.** 1965. The evolution of microbial diseases, p 20–36. *In* **Dubos RJ, Hirsch JG** (ed), *Bacterial and Mycotic Infections of Man,* 4th ed. Lippincott, Philadelphia, Pennsylvania.
3. **Dubos R.** 1956. Retrospectives and perspectives, pp. xv–xix. *In National Research Council, Proceedings, First International Conference on the Use of Antibiotics in Agriculture.* The National Academies Press, Washington, DC.
4. **Kossiakoff MG.** 1887. *De la propriété que possédent les microbes de s'accomoder aux milieux antiseptiques. Ann l'Inst Pasteur* **1**:465–476; see also: **Effront J.** 1891. *Koch's Jahresber Gärungorganisimen* **2**:154 (quoted in **Schnitzer RJ, Grunberg E.** 1957. *Drug Resistance of Microorganisms.* Academic Press, New York, New York, p 1); see also: **Davenport CB, Neal HV.** 1895. On the acclimatization of organisms to poisonous chemical substances. *Arch f Enrwicklungsmech Org* **6**:564–583.
5. **Ehrlich P.** 1907. Chemotherapie trypanosomen-studien. *Berl Klin Wochenschr* **44**:233–238.
6. **Morgenroth J, Levy R.** 1911. Chemotherapie der pneumokokkeninfektion. *Berl Klin Wochenschr* **48**:1560.
7. **Ehrlich P.** 1913. Chemotherapy: scientific principles, methods, and results. *Lancet* **2**:445–451.
8. See Reference 4.
9. See Reference 4.
10. **Akatsu S, Noguchi H.** 1917. The drug-fastness of spirochetes to arsenic, mercurial, and iodide compounds in vitro. *J Exp Med* **25**:349–362.
11. See Reference 7.
12. **Bryson V, Szybalski W.** 1952. Microbial selection. *Science* **116**:43–51.
13. **Tomes N.** 1998. *The Gospel of Germs: Men, Women and Microbes in American Life.* Harvard University Press, Cambridge, Mass.
14. **Summers WC.** 1991. From culture as organism to organism as cell: historical origins of bacterial genetics. *J Hist Biol* **24**:171–190.
15. **Lewis IM.** 1934. Bacterial variation with special reference to behavior of some mutabile strains of colon bacteria in synthetic media. *J Bacteriol* **28**:619–639.
16. **Dienert F.** 1900. Sur la fermentation du galactose et sur l'accountumance de levures à ce sucre. *Ann Inst Pasteur (Paris)* **14**:139–189.
17. **Karström H.** 1930. *Über die enzymbildung in Bakterien und über einige physiologische Eigenschaften der untersuchten Bakterienarten.* Doctoral dissertation. Universität Helsingfors, Helsinki, Finland.
18. **Stephenson M.** 1930. *Bacterial Metabolism.* Longmans, Green and Co, London, UK.
19. See Reference 16.
20. **Fildes P.** 1971. André Lwoff: An appreciation. *In* **Monod J, Borek E** (ed), *Of Microbes and Life.* Columbia University Press, New York, New York.
21. **Fildes P, Whitaker K.** 1948. Training or mutation of bacteria. *Br J Exp Pathol* **29**:240–248.
22. See Reference 20.
23. **Maclean IH, Rogers KB, Fleming A.** 1939. M. & B. 693 and pneumococci. *Lancet* **I**:562–568; see also **Abraham EP, Chain E, Fletcher CM, Gardner AD, Heatley NG, Jennings MA, Florey HW.** 1941. Further observations on penicillin. *Lancet* **ii**:177–189; and **Murray R, Kilham L, Wilcox C, Finland M.** 1946. Development of streptomycin resistance of Gram-negative bacilli in vitro and during treatment. *Proc Soc Exp Biol Med* **63**:470–474.

24. **Lederberg J, Lederberg EM**. 1952. Replica plating and indirect selection of bacterial mutants. *J Bacteriol* **63**:399–406.

25. **Brock TD**. 1991. *The Emergence of Bacterial Genetics*. Cold Spring Harbor Laboratory Press, Cold Spring Harbor, New York.

26. **Davis BD**. 1952. Bacterial genetics and drug resistance. *Public Health Rep* **67**:376–379.

27. **Luria SE, Delbrück M**. 1943. Mutations of bacteria from virus sensitivity to virus resistance. *Genetics* **28**:491–511.

28. See Reference 23.

29. **Newcombe HB**. 1949. Origin of bacterial variants. *Nature* **164**:150–151.

30. **Szybalski W**. 1956. Theoretical basis of multiple chemotherapy. *Tuberculology* **151**:82–85; see also Reference 25.

31. **Akiba T, Koyama K, Ishiki Y, Kimura S, and Fukushima T**. 1960. On the mechanism of the development of multiple-drug-resistant clones of *Shigella*. *Japan J Microbiol* **4**:210–222; see also: **Ochiai K, Yamanaka T, Kimura K, Sawada O**. 1959. Inheritance of drug resistance (and its transfer) between *Shigella* strains and between *Shigella* and *E. coli* strains, (in Japanese). *Nihon Iji Shimpo* **1861**:34–46; and **Watanabe T**. 1963. Infective heredity of multiple drug resistance in bacteria. *Bacteriol Rev* **27**:87–115.

32. **Summers WC**. 2015. Plasmids: histories of a concept, p 179–190. *In* **Gontier N** (ed), *Reticulate Evolution: Symbiogenesis, Lateral Gene Transfer, Hybridization and Infectious Heredity*. Springer, Berlin, Heidelberg.

33. **Lipsitch M, Singer RS, Levin BR**. 2002. Antibiotics in agriculture: when is it time to close the barn door? *Proc Natl Acad Sci USA* **99**:5752–5754; see also **Ghosh S, LaPara TM**. 2007. The effects of subtherapeutic antibiotic use in farm animals on the proliferation and persistence of antibiotic resistance among soil bacteria. *ISME J* **1**:191–203.

34. **Wellcome Explainer**. 17 November 2023. Why is it so hard to develop new antibiotics? https://wellcome.org/news/why-is-it-so-hard-develop-new-antibiotics (Accessed 7 October 2022.)

35. **Szybalski W, Bryson V**. 1952. Genetic studies on microbial cross resistance to toxic agents. I. Cross resistance of Escherichia coli to fifteen antibiotics. *J Bacteriol* **64**:489–499.

5 The Terrible Decade

Fort Bragg, North Carolina

"In the summer of 1940 an epidemic of an unusual febrile illness designated "Brushy Creek fever" affected about 35 residents of Wrens, Georgia (ref). During the summer of 1942, another epidemic of unusual febrile illness occurred at Fort Bragg, North Carolina (ref); there were 40 cases among soldiers in a localized area of the military base. After extensive laboratory studies no diagnosis other than "pretibial fever" (due to a rash over the pretibial area) or "Fort Bragg fever" was reported at that time, although reference was made to "what may be the same disease" in Wrens, Georgia. In the summer of 1943, the disease reappeared at Fort Bragg in seven soldiers quartered in the same area as the cases from the preceding year. Lieutenant Hugh Tatlock injected defibrinated blood from the seventh case intraperitoneally into guinea pigs and recovered a rickettsia-like organism from three of five animals (ref). . . .Tatlock could not show that this agent was responsible for the Fort Bragg fevers but published his findings "to record the known characteristics of the organism in the hope that future developments may clarify its significance" (1).

This retrospective account of mid-twentieth century microbe hunting succinctly captures the state of affairs from the front lines of infectious disease researchers. Outbreaks of "clusters" or sometimes even an individual oddity provided opportunities to hunt for the microbial cause of the disease, only to end with the confusing or inconclusive findings being filed away in "case reports" with the hope that

Magic Bullets, Miracle Drugs, and Microbiologists: A History of the Microbiome and Metagenomics, First Edition. William C. Summers.

this account would, in some way, be helpful to future microbe hunters who might stumble across an obscure but relevant case report. Microbial isolates such as that discovered by Tatlock in 1943 were checked against the gold standard of the latest edition of *Bergey's Manual*, and if it was not listed, it was assumed to be of marginal interest, perhaps a unique infection, likely never to be seen again, or in any event, not sufficiently exceptional to warrant much further concern. The analytical methods available to Tatlock and his colleagues were based on very limited physical characterizations of the isolates, the chemical and physical signature, if you will. Their tools were primarily biological characters: growth requirements that were only indirect indicators of the biochemical and physiological properties of the microbe. Did it grow well on this kind of nutrient medium and poorly on that kind? Did it require air (i.e., oxygen) to grow? Did it produce acid or carbon dioxide when it grew? In the few decades earlier, some powerful techniques to classify the surface molecules on microbes had been devised. Antibodies were created that could distinguish, by their binding, some of these surface proteins and carbohydrates that were characteristic of a given species or grouping of microbes. These immunological techniques revealed (and clarified) the diversity and complexity that was not apparent to earlier microbe hunters.

For the microbe hunters interested in viruses, the situation was even worse. Viruses, which are too small to even be seen under the regular light microscope, rely (almost) entirely on the cells they invade for their growth and metabolism, and were observable almost entirely by the diseases they caused in laboratory animals, chick embryos growing inside inoculated eggs, or isolated animal cells growing in artificial "tissue cultures." To this day, viruses resist our attempts at logical classification into a unified evolutionary scheme such as we have constructed for "higher" organisms.

A recent retrospective on this era by an esteemed American infectious disease leader, Anthony Fauci (1940–) captures the *zeitgeist* of this period:

> "I completed my residency training in internal medicine in 1968 and decided to undertake a 3-year combined fellowship in infectious diseases and clinical immunology at NIAID. Unbeknownst to me as a young physician, certain scholars and pundits in the 1960s were opining that with the advent of highly effective vaccines for many childhood diseases and a growing array of antibiotics, the threat of infectious diseases — and perhaps, with it, the need for infectious-disease specialists — was fast disappearing. Despite my passion for the field I was entering, I might have reconsidered my choice of a subspecialty had I known of this skepticism about the discipline's future" (2).

At that time, an eminent leader in American medicine, Robert Petersdorf (1926–2006) noted in the influential *New England Journal of Medicine* with regards to the number of physicians training in internal medicine, "Even with my great personal loyalties to infectious disease, I cannot conceive a need for 309 more infectious-disease experts unless they spend their time culturing each other" (3).

The theme of this chapter is the confluence of events, scientific, medical, and epidemiological, in the decade of the 1970s which fundamentally changed how we understand the microbial world and exposed the misplaced optimism of the previous decade.

In this decade, plus and minus a couple of years in each direction, major challenges arose to the core belief that we had a good catalog of the microbes in our environment, and that it was within our capacity to tame them for the betterment of our world and our species. We will review four such challenges to see how their power to upend our naïve microbial optimism set the stage for later science and new views of the microbial world. These four examples are only a few of the many that were accumulating in the medical and scientific journals at that time, but their social impact, their attention by the popular press, and their direct illustration of our microbiological hubris makes them good case studies. Legionnaire's disease, Lyme disease, and AIDS are now household words, while Lassa Fever may be less well-known, but it was a forerunner to the now infamous Ebola virus hemorrhagic fever. All four of these infections have caused serious human illnesses with high mortality rates. All four burst upon us without apparent warning with a lethal vengeance, and all four turned out to be caused by totally unsuspected microbes (in three of the four cases, microbes previously unknown to science). Microbiologists realized that they could no longer boast that "acute infectious diseases had been vanquished" and that infectious disease medicine was in its final "mopping up" phase. Indeed, this lesson has been reinforced most recently by the startling arrival of the novel COVID-19 virus and the surprised reaction of those who have forgotten this history and thus have been doomed to repeat it.

This period in American history seemed different. Old authorities were being challenged, the new medium of TV had become the standard for public information, U.S. involvement in the Vietnam war could be legitimately challenged in ways not previously possible, and science itself was under attack for its unforeseen consequences even as it produced new and amazing changes to our everyday lives. Against this turbulent background, both the public and the scientists held fast to the belief that infectious diseases were among our lesser worries. The Magic Bullets, the Wonder Drugs, and the increasingly technological state of medicine would protect us from the scourges of the past. To the surprise of most, however, the decade of the 1970s brought us new headlines of Mystery Illness and Deadly New Virus. The world suddenly became a dangerous and frightening place again. Where did all these "new" germs come from and why didn't our scientists know about them?

LASSA FEVER: 1969

In their widely admired and classic work on the natural history of infectious diseases, Nobelist Macfarlane Burnet (1899–1985) and his distinguished co-author, David O. White (1934–2004) provided the following epilogue to the fourth edition (1972) of their famous book:

> "On the basis of what has happened in the last thirty years, can we forecast any likely developments in the 70's? If for the present we retain a basic optimism and assume no major catastrophes occur and that any wars are kept at the 'brush fire' level, the most likely forecast about the future of infectious disease is that it will be very dull. There may be some wholly unexpected emergence of a new and dangerous infectious disease, but nothing of the sort has marked the last fifty years. There have been isolated outbreaks of fatal infections derived from exotic animals as in the instance of the laboratory workers struck down with the Marburg virus from African monkeys and the cases of severe haemorrhagic fever due to Lassa virus infection in Nigeria. Similar episodes will doubtless occur in the future but they will presumably be safely contained" (4).

This Lassa virus was to be an early warning that the "basic optimism" of Burnet and White was premature. Can we say that Lassa fever, as the human sickness that results from infection with this virus is called, is safely contained when in 2018 the Center for Disease Control reported that there are several hundred thousand cases annually with about 5000 deaths, and estimates that "In some areas of Sierra Leone and Liberia, about 10-16% of people admitted to hospitals annually have Lassa fever" (5).

Lassa fever, the disease, was first recognized by international medical scientists in the village of Lassa, Nigeria, a small town located in the Northeast of Nigeria in what is now Borno State. Its recognition, study, and notoriety came when it caused the deaths of two foreign missionary nurses and a laboratory scientist working in the United States (6). Another laboratory infection of a famous microbiologist who was investigating this African outbreak and his near-death saga further heighten awareness of this "new" disease. This disease, named for the location of its first recognition, would capture the public imagination as its terrors and mystery were widely recounted in the American media, in several dramatic books, and in the medical literature. Public and scientific awareness of the scope and dangers of infectious diseases, until then thought to be effectively vanquished, burst forth in the decade of the 1970s with a vengeance.

In January 1969 an American nurse, Laura Wine (1899–1969), working in the rudimentary missionary hospital in Lassa village took sick. Initially, she showed vague symptoms often experienced by inhabitants of this region of the West Sudan, perhaps a touch of malaria, or maybe one of the troublesome ills that she and her fellow missionaries had come to accept as part of their life in Africa. Quickly, however, she developed serious signs and symptoms of impending circulatory collapse, so the lone physician in Lassa, John Hamer (1923–2019), arranged, with much difficulty, to have Wine airlifted from a nearby landing strip to a regional hospital in the city of Jos, Nigeria where more care and resources might be available. Wine arrived at Jos nearly moribund, and it soon became clear that she had some sort of hemorrhagic fever, probably of viral origin. Despite medical heroics in the hospital at Jos, Laura Wine died a few days later. An autopsy performed there showed massive internal hemorrhages, but no findings that pointed to any clearly recognized cause. Samples of blood and tissues were, however, shipped to the U.S. for further study. A mystery disease but one of many that were known to those who tended to the medical needs of rural African populations.

No doubt it was what happened next which called attention to this particular illness from Lassa. Eight days after Wine's death, another case was observed, this time in a nurse, Charlotte Shaw (1923–1969), who had attended Nurse Wine in the hospital in Jos. Again, despite the full attention of the medical staff at the Jos hospital, Shaw died eleven days after the onset of the same symptoms that Wine had exhibited. Soon a third case appeared: Lily (Penny) Pinneo (1917–2012), a nurse and friend who had cared for Charlotte Shaw, was taken ill. By this time the physicians in Jos agreed that they were not equipped to deal with this apparently new disease, so Pinneo was, again after much difficulty, airlifted from Nigeria to Columbia Presbyterian Hospital in New York City (7).

The story of Lassa fever demonstrates once again, the role of contingency in human events (8). Some years before, a few physicians and scientists began to speculate what might be the role of the "shrinking world" of rapid air travel in the spread and recognition of infectious diseases. No longer could populations far apart count on geographic isolation to act as a *de facto cordon sanitaire* to protect them from diseases far and wide. Two of these forward-thinking physicians were John Frame (1918–2008) of Columbia University and Wilbur Downs (1913–1991), of the Yale Arborvirus Research Unit (YARU), organized and funded by the Rockefeller Foundation. They set up a collaboration to collect samples from foreign missionaries for analysis at the Yale labs to search for, characterize, and catalog known and unknown infectious agents. For several years, this project limped along, collecting samples more or less randomly when the occasional returning missionary was available. The case of Penny Pinneo, however, dropped into their lap, so to speak, in the right place at the right time. Several interested epidemiologists and laboratory virologists who had just such a problem on their minds set to work to better understand this apparently new disease from Nigeria. Frame and Downs, together with colleagues from their respective institutions, Columbia and

Yale, treated Pinneo with supportive care while carrying out detailed laboratory study of her illness. Fortunately, Penny Pinneo survived and later was able to provide both her immune serum and skilled nursing help in a future African outbreak of Lassa fever. At this point there were three known cases with two deaths from this mystery illness from Lassa village. The clinical course of the illness and its apparent transmission by personal exposure suggested that a new, probably unknown, virus was the cause of this lethal disease.

The laboratory at Yale was well-equipped with the then standard methods for detecting and studying viral diseases. These methods involved testing samples from a sick or deceased patient–blood, urine, feces, saliva, bits of tissue–by injecting them into several species of lab animals–usually mice, rats, guinea pigs, and sometimes rabbits–literally, "to see what happens." In addition, these samples were also overlayed on cultures of animal cells that had been growing in monolayers in Petri dishes or flat flasks. Again, the virologists looked at the cells under the microscope at intervals to see if anything (such as a virus) caused any observable changes in the appearance of the cell monolayers. These changes could be quite variable, but their characteristic appearance sometimes gave clues as to the identity of the suspected virus. The jargon for such changes is the cytopathic effect. Once one observed some effect on the animal or cell cultures due to something in the patient samples, there were a few other techniques that could be applied to further characterize and identify the virus. One more recent approach was to prepare the infected material (animal tissue or cultured cells) and observe them with the high magnification of the electron microscope. At very high magnification, the minute viruses can be observed and their structure, size, shape, etc. can be seen. Another technique was based on the use of immunology: the serum from a patient with the disease would be expected to react with the suspect virus whereas the serum from a normal individual would not. This reaction could be detected in several standard ways. This immunological approach was strong evidence linking the virus in the culture dish to the illness of the patient.

Fortunately, these standard laboratory methods were successful in identifying the Lassa fever virus in the Yale laboratory in straight-forward way. Science worked as expected and a new virus was discovered. Discovering a new virus, however, does not mean that the virus itself was new, just that the microbe hunters had been previously ignorant of it. It is likely that Lassa fever virus would take its place in a long list of exotic viruses that were accumulating in the freezers of microbe hunters around the globe except for two significant events: the laboratory infection of two Yale scientists, Jordi Casals (1911–2004), who recovered after being gravely ill, and Juan Román (1919–1969) who tragically died (9).

The infection of careful laboratory scientists, at the famous YARU, with a virus with a near fifty percent mortality rate frightened the scientific community, the

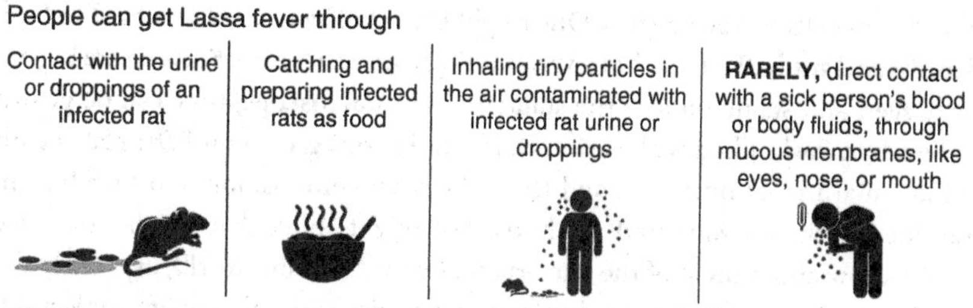

People can get Lassa fever through

Contact with the urine or droppings of an infected rat	Catching and preparing infected rats as food	Inhaling tiny particles in the air contaminated with infected rat urine or droppings	**RARELY,** direct contact with a sick person's blood or body fluids, through mucous membranes, like eyes, nose, or mouth

FIGURE 5.1 *Public health poster to educate the population in Africa about the rat reservoir of the Lassa fever virus (43).*

Yale community, and resulted in the termination of all live Lassa virus research at Yale and transfer of all potentially infectious samples to the CDC in Atlanta where new containment labs had just been constructed. The danger of this new "mystery virus" was broadcast to the world in a page one article in the *New York Times* on 10 February 1970: "New Fever Virus So Deadly that Research Halts" by the *Times'* medical reporter, Lawrence K. Altman (1937 –), an infectious disease physician with close ties to the Yale department of epidemiology where the Lassa virus work was being done (10).

As subsequent studies of Lassa fever and its causal viral agent showed, this new disease was not new, it was not rare, but it was hiding from the virus hunters in plain sight. Shortly after the initially recognized cluster of patients in Lassa village, several more outbreaks with more cases and still the appallingly high mortality were observed in West Africa. Eventually, the natural host or hiding place of the virus, the non-human reservoir of the virus, was identified as the common African rat (*Mastomys natalensis*) (Figure 5.1). Large studies have shown that the virus is endemic in central Africa. In Nigeria the prevalence of antibodies in the population was reported to be 21%, one in five residents showed evidence of prior infection with Lassa virus (11). Similar high prevalence rates were noted in Sierra Leone, Liberia, and Guinea. The CDC estimated in 2016 that 100,000 to 300,000 infections of Lassa fever occur annually, with about 5,000 deaths. How did virus hunters miss such an obvious and potent virus? How many similar microbes are "out there" but without the dramatic public relations exposure that just happened to land on Lassa virus? Lassa was just one early step in exposing the microbiological hubris that would erupt in the 1970s and change our understanding of the microbial world forever.

LEGIONNAIRES DISEASE: 1976

Lassa fever was caused by a virus, invisible under the standard light microscope, detectable only by its effects on inoculated animals that might, by good luck, be susceptible to the human virus, or again by chance make cytopathic lesions on cultured cells in Petri dishes. One might expect viruses would be hard to find, identify, and study. Bacteria, however, should be an easier case. Bacteria are bigger, easily seen and stained under the standard light microscope, they can be grown in the lab in flasks filled with various nutrient broths, grown on Petri plates with various mixtures of nutrients, and their chemical composition and metabolism examined with time-honored methods. We might think that by this time we should know about most of the bacteria that co-inhabit our world, especially the troublesome bacteria that cause disease or otherwise make human life unpleasant. Again, the decade of the 1970s was when we lost our innocence in bacteriology as well as in virology.

It was July 1976 in Philadelphia, in the city celebrated as the hallowed ground of the nation's founding, and two hundred years later some of the most fervent celebrants of that event gathered for their big convention. More than 2000 members of the various local Pennsylvania chapters of the American Legion, a group of retired U.S. military veterans, gathered at the Bellevue Stratford Hotel in downtown Philadelphia to celebrate the U.S. bicentennial with their annual State Convention. In the days just after the convention, which was held 21–24 July 1976, Sidney Franklin, a physician at the Philadelphia Veterans Hospital, saw several patients from the convention who presented with atypical pneumonia, poorly responsive to the usual treatments, and for whom the available laboratory diagnostic tests were inconclusive. By 2 August two of his patients had died. At this point, Franklin turned to the U.S. Center for Disease Control (CDC) in Atlanta for consultation, advice, and help. The CDC investigators "rounded up the usual suspects" and by 7 August they reported that they had "provisionally ruled out known bacteria, viruses and fungi" as related to the outbreak. By 15 August the investigators had assembled 182 cases from Legionnaires who had attended the convention with 27 deaths.

Again, the contingencies of history are informative. The bicentennial, the nation's birthday party, the convention of military veterans, not just ordinary citizens, and third, the widespread concern about a new influenza variant that had just ravaged a military installation at Fort Dix, New Jersey, prompting a near panic over a possible epidemic of a new "swine flu." And at that moment, a new "deadly, mystery disease" hit the headlines. "Legion Disease: Tracking a Killer" according to Science News, and "Mystery of the Killer Fever" on the cover of Newsweek in mid-August 1976, while the Time magazine cover story at the same time heralded "Disease Detectives: Tracing the Philly Killer" (12).

The hunt for the cause of the "Killer Fever" which seemed to specifically target elderly veterans partying at their convention would last for six months, take many detours, capture the public's attention, and become the whipping boy of the usual congressional grand-standers. Why do we spend so much of the peoples' money for the CDC, a government agency, that cannot tell us the cause of this lethal disease? Where are the microbe hunters?

All-in-all, this outbreak of what came to be called Legionnaires' Disease claimed 29 lives from among about 200 reported cases. While each life lost was certainly a tragedy, the overall annualized mortality of this cluster of cases was 2,000 times less that of influenza, a disease that arouses only minor concern among the American population. Instead, the Philly Killer became a center of prime-time attention. Between its first report of the outbreak on 3 August 1976 and the report of the discovery of the Legionnaires' Disease bacterium in mid-January 1977, the august *New York Times* published over 100 articles devoted to the Philadelphia outbreak. There were several Congressional hearings on the problem, and many speculative communications in the medical and lay literature. Everyone, it seems, had something to say about the mystery fever and its possible cause. The early failure of the CDC and others to identify a microbial agent that might be the cause of this type of lethal pneumonia gave way to rampant theorizing and fanciful proposals. Early on toxicological theories prevailed: heavy metal poisoning: cadmium from the coating on canned lemonade; rat poisons; chemicals used in cleaning the hotel rooms; nickel carbonyl to mention a few. One theory blamed emanation from "grates in the sidewalks, trees, subways entrances and median strips in the road near the Bellevue Stratford—any of these openings or objects could be the source of toxic substances that could make people sick who stood near it for a long period of time, say while watching a parade, but leave others unaffected if they were just passing by or waiting for a bus" (13). The CDC hotline was inundated with tips and suggestions: tainted mustard from a nearby hot-dog stand, and the presence of a hippie with a backpack in front of the hotel were just a few of the many proffered ideas.

A Congressional hearing took place in late November 1976 to inform the public on progress (or lack of it) by the microbe hunters (14). The hearing ranged over the new developments in biology such as the recombinant DNA work and the attendant public furor surrounding it in some quarters. Michael Crichton's 1969 book, *The Andromeda Strain*, came up for discussion. One witness reported that someone from the upcoming Republican Convention had inquired about protection against nickel carbonyl poisoning. One Legionnaire who spoke at the hearings said that he and his wife saw a suspicious-looking man at the hotel and testified: "As I looked around, I stood on the steps, there was a man in a bright, royal blue suit, single breasted, light brown hair, straight, parted on the right side and combed down toward his left eye. Head a receding forehead, light skin, about 38 to 45 years of age, approximately 5'10" to 6' tall and had a thick lower lip. This man was mingling among the crowd on the sidewalk and appeared to be saying things directed at the Legionnaires. For example, he said, 'It is too late now, you will not be saved.' Later I was told by another Legionnaire that he also said Legionnaires are doomed or something to that effect" (15).

As Leonard Bachman (1925–), secretary of the Pennsylvania Department of Health, stated in the Congressional hearings, "As a people we are not acclimated to failure" (16). Americans had come to believe in their scientific prominence in the world following the moon landing, the eradication of polio, and scores of Nobel prizes. Being accustomed to success, the existence of a mystery came as a shock. The fact that the mystery was a disease of unknown origin, a silent and invisible killer, magnified the anxiety. [...]

> "This anxiety carried over to the biological sciences with the revolution in molecular biology and the possibility of an Andromeda strain coming from the laboratory. In an early press conference, Dr. Bachman was asked if he had questioned research labs in the area about any recombinant DNA work that was in progress. The other dimension to the fear was the sudden realization of vulnerability. Society had begun to accept the idea of infallibility and constant progress in science. It was understandable with the eradication of countless diseases in the twentieth century that many were shocked that scientists could not quickly solve this problem" (Lauring A. 1995. Science under the microscope: the public and scientific investigation of legionnaires' disease. Unpublished manuscript).

Probably the explanation that gained most attention at this time was based on the known toxicity of a compound of the heavy metal, nickel, called nickel carbonyl. It was a known toxin that produced an illness that, in many ways, seemed to parallel that of Legionnaires' Disease. This theory had two vocal scientific advocates as well as a political ally in Representative John M. Murphy (1926–2015) (D-NY), chair of the House subcommittee which had held the November hearings. F. William Sunderman, Sr. (1898–2003), a respected toxicologist, and his son, F. William Sunderman, Jr. (1931–2011), a distinguished pathologist were strong advocates for this theory, especially after the presence of nickel was detected in many of the pathological samples. Murphy was supportive because of his interest and influence in passing the Toxic Substances Control Act of 1976.

The prevailing view, in the midst of this puzzling outbreak is accurately captured by statements by a scientist, David Sencer (1924–2003), the director of the CDC and, from a different point of view, by the editors of *The New York Times*. Sencer, in his opening sentence at the House committee hearing stated: "The outbreak of Legionnaires' disease which afflicted 180 people and caused 29 deaths has presented a number of unusual and complex features which I will try to describe. It has run counter to our expectation that contemporary science is infallible and can solve all the problems that we confront" (17). The journalists, too, reflected the dashed optimism of the scientists: days after the outbreak *The New York Times* wrote "Armed with the rich arsenal of scientific medicine, the investigators are fortunately well-equipped to solve the alarming puzzle, to deal with the threat of infectious disease and to prevent panic" (18). Just over a month later, the mood darkened: one "may hope that the initial errors made in studying Legionnaires' disease will not be repeated in future investigations of outbreaks of mysterious illnesses. The CDC has not added to the luster of its record by its performance here" (19).

The mystery surrounding Legionnaires' Disease even spread to the pop culture of the day with a song written by the rock icon, Bob Dylan (1941–), gently spoofing the situation:

> "*Some say it was radiation, some say there was acid on the microphone*
> *Some say a combination that turned their hearts to stone*
> *But whatever it was, it drove them to their knees*
> *Oh, Legionnaire's disease*"

In spite of having "ruled out" infectious diseases, especially the dreaded, anticipated epidemic of "swine flu," the CDC continued to consider some sort of infection as likely.

"The CDC announced that Joseph E. McDade (1941–) and Charles C. Shepard (1914 –1985) had isolated the bacterium now known as *Legionella pneumophila* on January 18, 1977. It was somewhat unexpected, "since as long ago as September, health officials had generally discarded the theory that an infectious disease was involved" (20). Because the bacterium was so small, it was not picked up in the initial review of the evidence. McDade returned to the lab during the "post-Christmas lull" because he had a haunting feeling that something was missed, described the discovery as similar to looking for a contact lens on a basketball four inches above the ground (21). It seems from these accounts that McDade's discovery was the "chance observation" of which Foege had spoken [William Foege, CDC director in testimony to Senate Committee in 1977]. Other accounts, however, stressed the patient investigation of the CDC (22). The CDC had been pursuing infectious agents through the fall of 1976 but it was not until McDade's observation that they had their breakthrough" (Lauring, unpublished).

The 168 days between the first hint of a new disease to the announcement of its microbial cause was both long and short. Both the scientific establishment and the public had become accustomed to near instant recognition, diagnosis, and successful treatment of infectious diseases. After all, what about microbiology lab tests, magic bullets, and wonder drugs? Five and a half months of uncertainty, suspense, anxiety, and fear seemed like a long time. On the other hand, in historical perspective, the discovery, identification, potential for treatment, and disease ecology in less than half a year seem pretty amazing. The microbe hunters came through, but had suffered a bit of overdue humility. As Adam Lauring aptly summarized this episode in medical history:

> "Because the CDC was under intense pressure from both the Congress and the general population, they turned to traditional methods in hopes of finding a quick answer. Unfortunately, the bacillus proved to be elusive and the discovery had to wait several months; the initial mis-direction pointed to the need for better toxicological analysis in the investigation. The Legionnaires' disease story represents a small Kuhnian crisis. Because of the infectious disease paradigm, the CDC devoted all of its time and resources toward finding a bacterium or virus. When none were discovered, the investigation reached a crisis situation and a possible paradigm shift. If an infectious agent was the cause, it did not fit into the standard categories" (Lauring, unpublished).

AIDS: 1981

AIDS, "Acquired Immune Deficiency Syndrome." The world now knows that this terrible affliction is caused by a particular infectious organism, a virus, now called human immunodeficiency virus, HIV. AIDS (at the time known by another name) was officially recognized as a human scourge in June 1981 with the publication of a case report of five individuals who had overwhelming infections with known microbes that seemed to be due to their lack of the usual immunity to these common microbes (23). In effect, they were sick with known microbes because they had a sickness of their immune system. Finding the cause of this sickness of the immune system would take nearly three more years, and in the end, it turned out to be caused by a totally new and unknown microbe, HIV (24). Why did it take so long and why was medical science so ignorant and astounded at this finding? Microbiological hubris again? Maybe.

To answer these questions it is crucial to look at the state of immunology, the science of the immune system, at that time. Immunity, especially in the medical context, was considered from two points of view: one was the role of the immune system in mitigating or preventing foreign materials, chemical or microbial, from harming the body. Since the early days of smallpox vaccination in the late eighteenth century and the advent of rabies and other immunizations in the late nineteenth and early twentieth century, immunology focused on the cells and serum proteins that mediated these wondrous phenomena. The other focus of immunology was the problem of allergy, that is, hypersensitivity to various substances to which the body's reaction was profound enough to affect health. . . skin rashes, asthma, drug sensitivities, and the like. Textbooks and medical journal articles in the 1970s had only a few vague references to rare genetic disorders in which there was a failure of the immune system to do its job (25). The new cancer drugs, along with radiation exposure, also were recognized as immune suppressants. Some of these agents were being exploited in a positive way to assist in the new science of tissue and organ transplantation by suppressing immune rejection of the foreign transplants. Clinicians often recognized certain physiological states such as malnu-trition and old age as leading to diminished immune competence. All-in-all, how-ever, immune deficiencies were a topic for esoteric, academic discussion; problems poorly understood and not easily dealt with.

As is usual, the investigation of the first causes of this apparently new disease of immune system failure focused on already known cases: exposure to chemicals such as occupational encounters with new products; use of illicit drugs new to the mar-ket; nutritional or metabolic disorders; something vaguely called "lifestyle" factors. These investigations followed the usual clinical and epidemiological approaches: look for known factors, look for common factors that might link the cases and, in some way, point to a common underlying cause. Consideration of a new, emergent microbe was quite far down the list, if present at all (26).

As is well-known, but of less direct relevance to this narrative, is the fact the epidemiological studies rapidly established linkages between the early cases and membership in several socially marginalized populations: gay men, injection drug users, certain groups of immigrants; and, increasingly, recipients of transfusions of blood and blood products, especially hemophiliacs who need treatment with clot-ting factor made from human blood pooled from a large number of donors. It was hard to find detailed, specific characteristics that were common to all these high-risk groups, however. The injection drug users and the hemophiliacs both were exposed to blood; the gay men and immigrants were not so universally exposed to blood. Epidemiology provides clues and points to possibilities but is short on direct evidence. Common exposures may suggest toxins; personal contacts can mean contagion; demographic associations may mean cultural or genetic factors are involved. For nearly two years, speculation ranged from the serious to the silly as to the "cause of AIDS."

One fact, however, provided a guiding light for medical researchers: a weak-ened immune system seemed to be the basis for the devastating inability to fight off usually innocuous infections. Whatever was the cause of AIDS, it was some-thing that hit the immune system, and something that seemed to be transmissi-ble. The microbe hunters had no ready candidate for such an agent. Again, as in the cases of both Lassa fever and Legionnaires' Disease, it seemed to be a mystery microbe. In the case of AIDS, however, there was one important clue. The immune defect seen in all the AIDS patients was a deficiency of the white blood cells that help fight infections, the T lymphocytes or T cells, for short (27). It was one of those fortunate coincidences that scientists searching for the causes of cancer had recently found that a certain rare type of cancer of the white blood cells, adult T-cell leukemia, most prevalent in Japan, seemed to be caused by a virus that was a member of a family of viruses known to cause cancer in experimental animals, viruses that have RNA as their genetic material. This virus, one of the first to be found to cause a human cancer, was called human T-cell leukemia virus or HTLV (28). Soon another related leukemia virus of this family was discovered and so the original HTLV was designated HTLV-1 and the later virus called HTLV-2.

Because these two HTLV isolates were the only known human viruses that infected human T lymphocytes, it seemed logical to test for their presence in the immune systems of AIDS patients. Perhaps a variant of HTLV had arisen that does not cause cancer in the T cells, but instead, simply kills them and in the process destroys the immune system, leading to AIDS. It is a well-known and widely analyzed story of how several research groups followed up this lead, sensing a potential Nobel Prize as well as providing the knowledge of a new microbe that might lead to both prevention and treatment of the growing AIDS scourge (29). Three research groups, two in the United States, and one in France, focusing on viruses that might kill or modify T lymphocytes, reported success in experiments using the classical approach of simply exposing the target T cells in laboratory cultures with material from sick patients, and then watching for some effect on the target cells. Again, as in the case of Lassa fever, this method would yield the answer: a new virus, confirmed by both electron microscope pictures and antibody tests, was found to be highly associated with the clinical immune deficiency seen in AIDS patients. This virus was named, after much quibbling, human immunodeficiency virus or HIV (30).

Where did HIV come from? Why didn't we know about it before 1982? Are there other viruses lurking "out there?" As in the other examples we have considered, both science and the public were unprepared for the surprises of this new virus. The public reaction was massive, not the least because of the political fodder AIDS and HIV supplied in the nascent culture wars that would engulf U.S. politics down to the present time. This new epidemic exposed the limits of existing knowledge, the limits of public health protections, and the limits of medical and scientific technologies. Once again, we were reminded of our microbiological hubris.

LYME ARTHRITIS: 1976

In the fall of 1975 "two worried mothers of afflicted children, acting separately, telephoned the Connecticut State Department of Health in Hartford. Within days their calls had set in motion a scientific detective process that turned the fears of 'hysterical mothers' into medical history" so wrote Boyce Rensberger (1943–), a science journalist at *The New York Times* in July 1976 (31). These concerned mothers were calling about a mysterious disease that seemed to have cropped up in and around the town of Lyme, Connecticut and was causing severe and debilitating joint and muscle pains in their children. Swollen and painful knee joints put kids in wheelchairs for weeks; leg pains crippled some even longer. The State Department of Health contacted the Yale rheumatology group for help in evaluating these cases of what appeared to be a kind of arthritis, inflammation of the joints.

In a classical epidemiological investigation of an apparent cluster of similar cases, the Yale team, led by a young physician, Allen Steere (1943–), and his mentor, Stephen Malawista (1934–2013), together with a young epidemiologist from the Connecticut Department of Health, David Snydman (1946–), located a cluster of over 50 cases in the three townships centered on the town of Lyme. Laboratory and clinical examinations as well as detailed personal histories of the afflicted individuals rather quickly indicated that something quite new was facing them (32). This was a new form of arthritis that seemed to be contagious, and based on analogies from other known instances, an arthropod vector such as a species of tick was their first suspect as part of the new puzzle. Since there were already several tick-borne viral diseases that might be related to the new diseases, christened "Lyme Arthritis" they focused their attention on this class of organisms. Ticks can cause diseases in two ways: humans can become allergic and have untoward reactions to the saliva, excreta, or other parts of the tick itself, and alternately, the tick may simply be the carrier of a microbe, a virus or bacterium, that was the true "cause" of the arthritis. Steere and Malawista were rheumatologists, immersed in the immunological complexities of human immune reactions that led to various diseases. They soon initiated collaborations with virologists to search for potential microbial origins of Lyme arthritis.

It did not take long for the field investigations to make two key observations. Ticks were prevalent in the area around Lyme, and there was a condition reported in Europe first in 1909, that involved ticks in what appeared to be a skin reaction called *erythema chronicum migrans* (ECM, "chronic migrating redness") (33). Only rare cases of ECM had been reported in North America, and its cause was debated, perhaps an allergic reaction to tick material, perhaps a microbe carried by the tick. The cure of ECM by penicillin was taken as evidence of a bacterial cause, but the efficacy of penicillin was challenged in several studies. ECM, although a murky clue, would provide a crucial hint pointing to the cause of this new form of arthritis in and around Lyme, Connecticut.

The initial search for an infectious cause of Lyme arthritis was optimistic: just as in the case of Lassa fever, scientists exposed animal cells cultured in bottles to patient samples, hoping to see some effects on the cells which would indicate a virus, known or unknown. Warren Andiman (1945–), a young virologist just starting his career at Yale exposed four types of cells to samples from patients with Lyme arthritis but saw nothing. Andiman was still optimistic: "It's conceivable that it's an organism we don't know of, but it doesn't seem likely." Steere and Malawista were upbeat as well, Rensberger reported: "They conclude only that they have not yet found the right method for recovering and growing the virus." "Even after the virus or microbe is found," Malawista said, "questions are likely to remain. Where did it come from? Is it a new mutant of a known disease virus? Is it a new variant of a virus that causes one of the other forms of arthritis whose cause was unknown" (34)?

The epidemiological and other, what might be called circumstantial, evidence suggested nearly from the first recognition of Lyme arthritis, that ticks might be the vector of a microbial cause of the disease. Medical entomology, the study of insects and arachnids (ticks, mites, spiders, and scorpions) is a rather esoteric discipline and research on these abundant organisms is limited to a few centers and institutes, one of which is the U.S. Public Health Service Rocky Mountain Laboratory in Hamilton, Montana. It is in this laboratory that much famous tick research has been done. This lab grew out of the pathbreaking work of Howard Ricketts (1871–1910), discoverer of the cause of the tick-borne disease that was prevalent in the area, Rocky Mountain Spotted Fever. In 1909 Ricketts identified, isolated, and characterized the bacterial cause, and in the process explored a previously murky group of bacteria, now named *Rickettsia* in his honor.

A second historical connection is relevant to this account of Lyme arthritis as well. At one time, an important medical sign of the venereal disease, syphilis, was the recognizable skin lesions, both on the genitals (chancre) and the rest of the skin (generalized rash on the palms, soles, and mucous membranes. These signs of this serious disease frequently brought the sufferer to consult a dermatologist. The result was a natural conjunction of the developing medical specialties of dermatology and venereology, a combination that often seems puzzling in the current medical landscape. When it was found, in 1905, that syphilis was caused by a spiral-shaped bacterium called *Treponema pallidum*, the dermatologists who saw syphilis patients became intimately acquainted with this microbe, a member of a group of spiral bacteria known as spirochetes. Since it was the dermatologists who saw patients with rashes of ECM as well as syphilis, it is likely that some might wonder if spirochetes might be involved in ECM as well as syphilis. Indeed, at least a few got this idea during the golden age of microbiology beginning after the end of WWII and the advent of the "wonder drugs." In 1948, Carl Lennhoff (1883–1963) working in Sweden reported seeing "elements" that looked like spirochetes in tissues from patients with ECM, among others (35). A year later, Sven Hellerström (1901–1977) described a case of ECM and in the discussion, he suggested that a spirochete was the cause of ECM: "Our investigations tend to demonstrate that a spirochete is the cause of this disease. Definite evidence is still lacking" (36).

This conjunction of ticks, dermatology, spirochetes, ECM, and Lyme arthritis came together in the person of Willy Burgdorfer (1925–2014) working on tick biology, who, together with his colleague, Jorge Benach (1945–) of the New York State Health Department, had been studying the ecology of Rocky Mountain spotted fever on Long Island (37). One collection of ticks of the genus *Ixodes* came from Shelter Island (part of Long Island), a place where Lyme arthritis had been found. *Ixodes* was already the suspect genus of tick implicated by epidemiological surveys as the potential vector for the Lyme agent, whatever it might be. When Burgdorfer examined the ticks from Shelter Island, he noted that two of the 44 ticks had peculiar tiny worms (microfiliariae) which he wanted to identify. He wanted to find out if these worms were also in the digestive system, not just the body fluid of the ticks. Under a microscope he dissected the gut of the tick and prepared stained microscope slides of the gut material to look for the worms. He found none. The worms did not inhabit the gut. To his surprise, however, the tick guts contained a strange appearing spirochete. Soon he had dissected and examined 124 ticks; sixty percent of them were infected with the spirochete. It was likely that by looking for microfilariae as a possible cause, instead, he had just identified the cause of ECM and Lyme arthritis as a new bacterium.

The bacterial group known as spirochetes are sensitive, finicky eaters so they are notoriously hard to grow in the laboratory. Special growth media, especially clean glassware, and long incubation times are needed. At the Rocky Mountain Laboratory Burgdorfer's colleague, Alan Barbour (1945–) undertook the challenge and was able to isolate and grow the spirochete that Burgdorfer had observed in so many of the ticks from Long Island where Lyme arthritis was endemic (38).

Once this new bacterium had been isolated in pure culture, it was possible to test for antibodies in ECM patients, in Lyme arthritis patients, and in control health individuals. These immunological tests all linked the new spirochete to the presence of these two clinical entities (39). The causal connection was further strengthened when it was found that the spirochete was sensitive to penicillin and other drugs that were curative when administered to patients with these conditions (40). In 1984 Burgdorfer's new spirochete was officially recognized as yet another mystery microbe successfully identified and entered in the pantheon of *Bergey's Manual* as *Borrelia burgdorferi* (41).

A DECADE IN REVIEW

The four previously unknown microbial diseases that we have examined in this chapter all emerged in the era of the 1970s, plus or minus a few years. Their serious impact promoted widespread fear, yet at the same time they gave new energy to microbiologists who had grown a bit complacent. And these four new microbes were not the only surprises of this period. Soon a microbe would be found to cause peptic ulcer disease, long thought to be a marker of successful executive stress. Drug-resistant, "flesh-eating bacteria" would hit the headlines. Toxic shock syndrome made the news in 1978, too, along with one of its main microbial causes, *Staphylococcus aureus,* the "golden staph" of the later MRSA (methicillin-resistant *Staphylococcus aureus*) emergence. In sum, this era was both challenging and frightening for microbiologists and infectious disease experts. In the next chapter we will see how this loss of microbial innocence would herald new views of the microbial world, stimulate new technologies as well as new thinking about microbes as our co-inhabitants of the planet Earth (42).

NOTES AND REFERENCES

1. **Hébert GA, Moss CW, McDougal LK, Bozeman FM, McKinney RM, Brenner DJ.** 1980. The rickettsia-like organisms TATLOCK (1943) and HEBA (1959): bacteria phenotypically similar to but genetically distinct from *Legionella pneumophila* and the WIGA bacterium. *Ann Intern Med* **92**:45–52

2. **Fauci AS.** 2022. It ain't over till it's over. . . but it's never over–emerging and reemerging infectious diseases. *N Engl J Med* **387**:2009–2011.

3. **Petersdorf RG.** 1978. The doctors' dilemma. *N Engl J Med* **299**:628–634.

4. **Burnet M, White DO.** 1972. p 263. *In Natural History of Infectious Diseases,* 4th ed. Cambridge University Press, Cambridge, UK.

5. **Centers for Disease Control and Prevention.** 2022. Lassa fever fact sheet. https://www.cdc.gov/vhf/lassa/pdf/lassa-factsheet-508.pdf. Accessed 15 December 2023.

6. **Frame JD, Baldwin JM Jr, Gocke DJ, Troup JM.** 1970. Lassa fever, a new virus disease of man from West Africa. I. Clinical description and pathological findings. *Am J Trop Med Hyg* **19**:670–676.

7. **Pinneo L, Pinneo R.** 1971. Mystery virus from Lassa. *Am J Nurs* **71**:1352–1355.

8. **Fuller JG.** 1974. *Fever! The Hunt for a New Killer Virus.* Reader's Digest Press, New York, New York.

9. See Reference 8. pp 143, 183.

10. **Altman LK.** 1970. New fever virus so deadly that research halts. *The New York Times* 10 February, pp 1, 25.

11. **Richmond JK, Baglole DJ.** 2003. Lassa fever: epidemiology, clinical features, and social consequences. *BMJ* **327**:1271–1275; see also **Agbonlahor DE, Akpede GO, Happi CT, Tomori O.** 2021. 52 years of Lassa fever outbreaks in Nigeria, 1969-2020: an epidemiologic analysis of the temporal and spatial trends. *Am J Trop Med Hyg* **105**:974–985.

12. **Anonymous.** 1976. Legion disease: tracking a killer. *Sci News* **111**(14 August):102; see also **Clark M, Clark E, Gosness M.** 1976. The mystery fever; hunt for a killer. *Newsweek* **88**(16 August):16–18, and **Anonymous.** 1976, The Philadelphia killer. *Time* **108**(16 August):64–68.

13. See Reference 11.

14. **Committee on Interstate and Foreign Commerce, US House of Representatives.** 1976. Legionnaires' Disease hearing before the subcommittee on consumer protection and finance, 94th Congress, Nov. 23-24. Serial 94–159. https://li.proquest.com/elhpdf/histcontext/HRG-1976-FCH-0035.pdf. Accessed 1 November 2022.

15. See Reference 14, p 82.

16. See Reference 14, p 26.

17. See Reference 14, p 93–94.

18. **Editor.** 1976. Philadelphia puzzle. *New York Times* August 5:30sec1.

19. **Editor.** 1976. 'Legionnaires' Disease.' *New York Times* September 15:44:sec1.
20. **Adams V, Ferrell T.** 1977. Legionnaires' Disease 'solved.' *New York Times* January 23:9sec1.
21. **Schmeck H Jr.** 1977. A haunting feeling about a clue led to Legionnaires' Disease key. *New York Times* January 24:25sec1.
22. **Anonymous.** 1977. Medicine: found: the Philly killer, perhaps. *Time* January 31:47.
23. **Marx JL.** 1982. New disease baffles medical community: 'AIDS' is a serious public health hazard, but may also provide insights into the workings of the immune system and the origin of cancer. *Science* **217**: 618–621; see also **Essex M.** 1986. The etiology of AIDS: introduction and overview. pp 3–7. *In* Kulstad R (ed), *AIDS: Papers from Science, 1982-1985.* AAAS, Washington, DC.
24. **Barré-Sinoussi F, Chermann JC, Rey F, Nugeyre MT, Chamaret S, Gruest J, Dauguet C, Axler-Blin C, Vézinet-Brun F, Rouzioux C, Rozenbaum W, Montagnier L.** 1983. Isolation of a T-lymphotropic retrovirus from a patient at risk for acquired immune deficiency syndrome (AIDS). *Science* **220**:868–871; see also **Levy JA, Hoffman AD, Kramer SM, Landis JA, Shimabukuro JM, Oshiro LS.** 1984. Isolation of lymphocytopathic retroviruses from San Francisco patients with AIDS. *Science* **225**:840–842.
25. See Reference 4. p 90; see also **Holvey DN (ed).** 1972. Immunologic deficiency diseases. pp 321–324. In *The Merck Manual of Diagnosis and Therapy.* Merck Sharpe & Dohme Research Laboratories, Rahway, New Jersey.
26. **Jaffe HW, Bregman DJ, Selik RM.** 1983. Acquired immune deficiency syndrome in the United States: the first 1,000 cases. *J Infect Dis* **148**:339–345.
27. There are two recognized types of lymphocytes, named for their site of origin in the body. Those made by the thymus, a lymphocyte-producing organ located just behind the breastbone, are called T-lymphocytes, and those made by the bone marrow are called B-lymphocytes (not for "bone marrow" but for the "bursa of Fabricius," a lymphoid organ in birds where they were first discovered). Both B and T cells have important, but distinct, roles to play in the immune system.
28. The nomenclature of this group of viruses has changed in recent years and is both confusing and complex. The original name HTLV was changed to "human T-lymphotropic virus" and the original HTLV-3 was renamed "human immunodeficiency virus" or HIV. Another isolate of a different virus is currently designated HTLV-3; see **Vallinoto AC, Rosadas C, Machado LF, Taylor GP, Ishak R.** 2022. HTLV: it is time to reach a consensus on its nomenclature. *Front Microbiol* **13**:896224.
29. **Shilts R.** 2007. *And the Band Played On: Politics, People, and the AIDS Epidemic, 20th-Anniversary Edition.* St. Martin's Griffin, New York, New York; see also **Kingman S, Steve Connor S.** 1989. *The Search for the Virus.* Penguin, Harmondsworth, UK.
30. **Marx JL.** 1986. AIDS virus has new name--perhaps. *Science* **232**:699–700.
31. **Rensberger B.** 18 July 1976. A new type of arthritis found in Lyme, p 1. *New York Times.* New York, NY.
32. **Steere AC, Malawista SE, Snydman DR, Shope RE, Andiman WA, Ross MR, Steele FM.** 1977. Lyme arthritis: an epidemic of oligoarticular arthritis in children and adults in three connecticut communities. *Arthritis Rheum* **20**:7–17.
33. **Afzelius A.** 1910. Erythema migrans. Verhandlungen der dermatologischen Gesellschaft zu Stockholm 1909. *Archiv f Derm u Syph* **101**:404; see also **Afzelius A.** 1921. Erythema chronicum migrans. *Acta Derm Venereol* **2**:120–125.
34. See Reference 30.
35. **Lennhoff C.** 1948. Spirochaetes in aetiologically obscure diseases. *Acta Derm Venereol* **28**:295–324.
36. Hellerström quoted in: **Burgdorfer W.** 1984. Discovery of the Lyme disease spirochete and its relation to tick vectors. *Yale J Biol Med* **57**:515–520; see also **Burgdorfer W, Barbour AG, Hayes SF, Benach JL, Grunwaldt E, Davis JP.** 1982. Lyme disease–a tick-borne spirochetosis? *Science* **216**:1317–1319.
37. Despite its name, Rocky Mountain spotted fever is quite prevalent in other regions, especially along the east coast of the United States. In the interval 1971-1976 there were a total of 124 cases of Rocky Mountain spotted fever with eight deaths on Long Island.

38. **Barbour AG.** 1984. Isolation and cultivation of Lyme disease spirochetes. *Yale J Biol Med* **57**:521–525.

39. **Steere AC, Grodzicki RL, Kornblatt AN, Craft JE, Barbour AG, Burgdorfer W, Schmid GP, Johnson E, Malawista SE.** 1983. The spirochetal etiology of Lyme disease. *N Engl J Med* **308**:733–740.

40. **Van Hout MC.** 2018. The controversies, challenges and complexities of Lyme disease: implications for medical education, clinical practice and research. *J Pharm Pharm Sci* **21**:429–436.

41. **Johnson RC, Schmid GP, Hyde FW, Steigerwalt AG, Brenner DJ.** 1984. *Borrelia burgdorferi* sp. nov.: etiologic agent of Lyme disease. *Int J Syst Evol Microbiol* **34**:496–497.

42. **Sencer DJ.** 1971. Emerging diseases of man and animals. *Ann Rev Microbiol* **25**:465–486; see also **Morse SS.** 2001. Factors in the emergence of infectious diseases, pp 8–26. In Price-Smith AT (ed). *Plagues and Politics. Global Issues Series.* Palgrave/Macmillan, London, UK.

43. **Centers for Disease Control and Prevention.** 2022. What you need to know about Lassa fever. https://www.cdc.gov/vhf/lassa/pdf/What-you-need-to-know-about-Lassa-508.pdf. Accessed 15 December 2023.

6 Loss of Innocence

While the 1970s experienced new and frightening outbreaks of infectious diseases, apparently new to science and medicine, outbreaks that for the first time started to suggest that our knowledge of the microbial world was not as robust as we had come to believe, new and exciting breakthroughs in the laboratories around the world were taking place at the same time. These breakthroughs provided new tools and approaches to detect, identify, and classify microbes by direct examination of the genes of these tiny organisms. Instead of characterizing microbes by their appearances under the microscope and their nutritional requirements, the classical methods for identifying and classifying microbes and the mainstay of *Bergey's Manual,* these new methods analyzed the genes that were the bases for these characteristics. The chemical structures of microbial genes succumbed to detailed analysis that provided insights only dreamed of a few decades earlier. Comparing genes by determining the sequence of chemical units (the nucleotides) in the DNA (initially through the RNA sequences which represent a direct copy of the DNA code) allowed scientists to discover genetic relationships between strains as well as unambiguously identify and distinguish any given isolated microbe (1).

There are two related concepts involved in the modern search for order and understanding of the microbial world. *Taxonomy* is the process by which things are identified, named, and organized. The other endeavor is *Phylogenetics* which tries to understand evolutionary relationships between individual species using hereditary features as the basis for such relationships. Prior to Darwin and the modern understanding of organic evolution, taxonomy was based on appearances,

Magic Bullets, Miracle Drugs, and Microbiologists: A History of the Microbiome and Metagenomics, First Edition. William C. Summers.

ecological factors, and analogies. We identified species by names such as the South African long-toed tree frog (place, appearance, habitat, analogy). These identifying features were only indirectly linked to heredity, linkages that only hinted at evolutionary explanations of origins and relationships.

Most biologists now agree that a useful and meaningful classification scheme for any group of organisms should be compatible with the principles of evolutionary history of the group of organisms (2). That is, the basis for classification should be the evolutionary, and hence genetic, relatedness of species. Such a scheme shows us how genetic changes over time could evolve from a primordial, ancestral organism to produce multiple different, but related, descendent genetic structures. In the early 1960s Emile Zuckerkandl (1922–2013) and Linus Pauling (1901–1994) examined the way DNA, RNA, and protein molecules could function as the chemical basis for molecular phylogeny. They concluded that DNA and RNA are the best tools (3).

RIBOSOMAL RNA

Methods for determining the gene sequences as represented in the RNA found in cells were devised in the mid-1960s primarily in the laboratory of Frederick Sanger (1918–2013) in Cambridge, England at the Medical Research Council's Laboratory of Molecular Biology (4). These methods were novel, technical, and expensive, but by the early 1970s, they gradually were adopted in several laboratories around the world. Gene sequencing was about to revolutionize genetics, and indeed, revolutionize biology as a whole.

For a few technical reasons, it was easier to determine the sequence of nucleotides in RNA than to do it for DNA. Because the RNA sequences are direct copies of the DNA sequences, biochemists tackled RNA sequences first. Since there are several billion nucleotides in each mammalian cell, usually strung together into a few chunks of many millions each, the chromosomes, the first task of the biochemists was to find out how to separate out one small bit of a gene in sufficient quantity to apply their new sequencing methods. Fortunately, they found right away that the cell had done some of this work for them. Every cell has an abundance of an RNA-containing granule that is used to synthesize proteins, needed by all cells. These granules, called ribosomes (RNA bodies) are easy to prepare in large quantity, and were found to contain three sizes of distinct RNA molecules. The cell had packaged a tiny subset of all the RNA copies of some DNA genes and presented them to the biochemists in the form they could deal with. The middle-sized fragment, found in many microbes, dubbed 16S rRNA based on its size as measured in certain "s" units, contains about 1,500 nucleotides in a linear linkage. This *genetic marker* became the first direct gene sequence to be employed to study the diversity

---

The Logic of 16S rRNA in Evolutionary Biology

All cells have ribosomes, the structures that synthesize proteins

↧

Genes that code the components of ribosomes evolved very early in the history of life

↧

All ribosomal genes have a common ancestor gene

↧

Comparisons of the ribosomal gene sequences show genetic diversity and evolutionary histories of organisms.

↧

The 16S rRNA molecule of the ribosome is easiest to sequence

FIGURE 6.1 *Logic for selecting the gene for 16S rRNA for microbial genealogy studies.*

of the microbial world in 1977 when George Fox (1945–), Kenneth Pechman, and Carl Woese (1928–2012) showed that by analyzing just tiny fragments of RNA from bacterial ribosomes, it was possible to classify and identify distinct bacterial species (5) (Figure 6.1). They noted several features of this approach to what they called "molecular systematics." First, because the ribosome is a complicated structure, and the rRNA is an integral part of it, there would be strong constraints on the possible changes over time in the structure of the rRNA. In gene-speak the sequences are "highly conserved;" that is, they should evolve slowly enough so that diversity between related species would not obscure their common genetic ancestry (6). Second, only small parts of the rRNA were needed to see relationships between related organisms; partial sequence data would be enough. The belief that ribosomes are of ancient origin (because all organisms seem to share the common feature of life based on proteins) suggested that rRNAs would be a marker for deep evolutionary probing. "An organism's genome seems to be the ultimate record of its evolutionary history" (7). For practical reasons they settled on 16S rRNA. This prediction was quickly realized. Later the same year, Woese and Fox went on to show how their "molecular systematics" revealed a completely new and unexpected grouping of living organisms, a group they called *archaebacteria*, or archaea, based on their presumed ancient evolutionary origin (8).

Woese and his collaborators, in the 1970s, introduced an entirely new concept into the field of microbiology: instead of two major phylogenetic groupings, usually called domains (sometimes kingdoms), the eukaryotes and the prokaryotes, based on the presence or absence of a distinct cell nucleus, they proposed a new classification based on three domains that could be distinguished by analysis of relationships of the 16S rRNA molecules. They examined sequences from a wide variety of both eukaryotic and prokaryotic organisms and then grouped them by the frequencies of common rRNA sequences. Instead of the expected two groupings, eukaryotic and prokaryotic, they found three groupings: the eukaryotic organisms fell into a common group, more or less as expected, but the prokaryotic organism comprised two quite distinct groups. Further, one of the prokaryotic groups contained only microbes that were known to produce methane gas as a product of their metabolism. These microbes, the methanogens, were relatively unknown and rarely studied, and were presumed to be of ancient origin because they seemed to be well-suited to the environment thought to exist on the primitive Earth. They get their energy from reduction of carbon dioxide to methane rather than by oxidation of organic carbon to carbon dioxide. Surprisingly, the rRNA analysis showed that the methanogens, although classically viewed as bacteria, were no closer to the "typical bacteria" than they were to the eukaryotes. Woese and Fox proposed a new grouping, represented at this point only by the methanogens, that is ancient, possibly the ancestors of other more "modern" microbes, the archaebacteria, or archaea. They proposed that the traditional category of prokaryote be rechristened as "eubacteria." This grouping of microbes from the 1970s has become widely accepted in the past nearly half-century as more and more support has accrued to it, especially in the form of newly discovered microbes living in diverse and extreme environments (Figure 6.2).

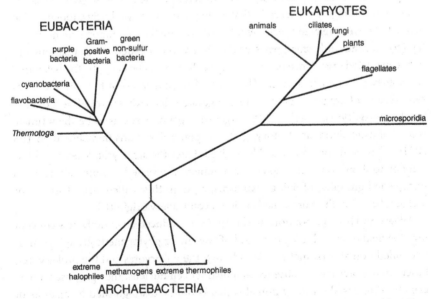

FIGURE 6.2 *Tree of life proposed by Woese in 1987 based on evolutionary relationships of the 16S rRNA gene (17).*

ARCHAEA AND EXTREMOPHILES

The new tool for characterizing microbes, the ribosomal RNA chemical sequences of nucleotides, revealed a surprising new approach to the evolutionary organization of these small forms of life. The archaea that Woese and his colleagues first described was a small set of known bacteria that produced methane as a byproduct of their energy metabolism. Such production of methane was long ago observed as "marsh gas," sometimes giving rise to fires in the night that were both mysterious and frightening. By the mid-twentieth century, however, marsh gas was well-understood as a microbial byproduct, but such organisms were considered a curiosity, not of interest as a pathogen or technologically useful organism. But the methanogens were not the only strange microbes that occasionally appeared in the scientific literature. Two other groups of bacteria puzzled the few biologists who considered bacteria as objects worth study beyond either their commercial utility or medical relevance. Environments with very high salt concentrations such as salt pans (pool for evaporating saltwater to recover salt crystals) or foods or other materials treated with concentrated salt as a preservative sometimes can become contaminated with bacteria that will only grow in the lab if they are given a high salt culture medium. These odd organisms were dubbed "halophiles," salt-lovers. These halophiles really love salt. . . not just like sea-saltiness, but a solution of salt that was so concentrated that no more salt can be dissolved in it. Most known bacteria are killed by so much salt. How do halophiles manage to live in such environments?

A second group of unusual bacteria were found in hot environments, not just warm, but hot, such as thermal hot springs. Again, these "thermophiles," heat-loving bacteria, grow only at temperatures that kill most "normal" bacteria. There were even a few bacteria, isolated from cooling water from nuclear reactors, that are many, many times more resistant to x-radiation than most bacteria.

The interest in new groups of bacteria took on new life in the 1970s and 1980s, partially motivated by the fact that several of the known examples of halophiles and thermophiles turned out to belong to the archaea when their 16S rRNA was analyzed. Collectively, these organisms came to be known as "extremophiles." These organisms provided some basic challenges to scientists. First, of course, was the problem of laboratory growth, finding the optimal conditions to propagate the bacteria in the laboratory for study. More interesting, however, was the question of how they could live in conditions that inactivated the key cellular molecules that make life possible for the known, "normal" bacteria. Very high salt coagulates proteins, inhibits intermolecular interactions, and blocks enzyme activities. High temperatures destroy the structures of both DNA and proteins as well as cause the break-up of many small molecules in the cell that are essential for growth. How have the extremophiles evolved ways to solve such problems? Soon, it became important to know just how widespread these phenomena were, and how they evolved. More examples of diverse organisms might help. Bioprospecting became respectable.

Thomas Brock (1926–2021) was a microbiologist with broad interests, which included extremophiles. In the late 1960s, he started collecting samples from the hot geysers in Yellowstone National Park in Wyoming. Many of the beautiful geysers in Yellowstone owe their colored appearance to various pigmented algae and bacteria that thrive in the mineral-rich waters. These geyser pools, however, are often steaming hot, a potential home for thermophiles. Brock made detailed bacteriological studies of this environment and found diverse populations of thermophiles, as anticipated. One of his isolates, *Thermus aquaticus*, grows optimally at 65–70°C, but can still live at 80°C. This organism became, seemingly fortuitously, famous and well-studied because it was soon discovered that its DNA polymerase was stable even at 95°C, making it an ideal biological reagent in automated polymerase chain reaction (PCR) devices (9).

The halobacteria also have generated special interests. Many of the halobacteria have a reddish-purple pigment called *bacteriorhodopsin*, which traps light which the bacteria use to provide energy to the cell. In a remarkable bit of evolutionary mystery, this protein turns out to be structurally related to the rhodopsin pigment in the retina of the vertebrate eye, the molecule that allows us to detect light. Like many halophile proteins, bacteriorhodopsin is more stable in the laboratory than the delicate rhodopsin molecule, a boon to the crystallographers studying the mechanisms of vision (10).

The concept of the extremophiles, of course, is not a morphological or genetic characterization, but rather an ecological description. As such it did not add directly to our understanding of the "tree of life" or the organization of the microbial world, but it did stimulate searches and serious thought about microbes that might have been overlooked simply because microbiologists failed to look where such unknown microbes might lurk. Soon microbes were found thriving around deep-sea vents where the temperatures and pressures were so high that they were nearly impossible to provide in the laboratory setting. Moon rocks were studied for space microbes (without much concrete success, however). Microbes that grow without oxygen, microbes that grow with only nitrogen, and microbes that grow in the cold, below the freezing point of water (psychrophiles), all began to appear in scientific reports describing growth in "the outer reaches of life" in the words of the British microbiologist, John Postgate (1922–2014) (11).

POPULATION SIZE AND DIVERSITY

With the examples of the archaea and the extremophiles, microbiologists began to look seriously into two related questions: "how many microbes are there in the world?" and "how many kinds of bacteria exist?" Questions of population size and population diversity. The diversity problem had been lurking around the margins of microbial science from at least the 1930s. The usual way to study bacterial diversity was to spread a sample taken from nature, a pathological lesion, or some other specific ecological niche on an agar culture plate in sufficient dilution so that each microbe had room to multiply and form a visible isolated colony. Such colonies, known as clones, or often, pure cultures, were assumed to arise from single "founder" microbes and hence represent all the progeny of that microbe carrying the characteristics of the founder type. Colonies were differentiated by their shape, color, size, and other visible markers. Sometimes they could be analyzed for additional chemical markers such as staining with dyes, reactions with antibodies, or metabolic activities. These identifying characters had been sufficient to produce the large list of bacterial species that found their way into *Bergey's Manual*. Still, these characteristics of diversity were a limited subset of ways one might want to distinguish various bacterial tribes. Simple morphology on a culture plate was not enough to reliably distinguish the various kinds of common bacteria that normally inhabit the human intestine, for example. Often a colony was identified only as a "coliform," that is, it looked like the well-known *Escherichia coli*, but might be another closely related organism. Further lab tests were often needed.

In addition to imprecise tools to identify colonies, there was the fundamental problem of numbers. A standard culture plate (Petri dish) is about 3-1/2 inches in diameter and even with very small colonies, several hundred colonies is about the maximum number that can grow distinctly and separately on such a culture plate. It is easy to recognize the few very dominant species, but if the sample contains minority species, say less than 1 in 1000, it becomes hard to recognize among the crowded colonies of the dominant species. Scientists just could not see rare, but possibly important minor species when they relied on agar plate cultures. Some attempts to search for minor species that might have unusual growth requirements were attempted. Culture conditions that inhibited the growth of the dominant forms allowed identification of rare species in some cases, but systematic studies of species diversity in microbiology was late to the lab when compared to the long history of bioexploration, collecting, and cataloging larger organisms.

By the late 1960s and on into the 1980s, microbiologists started to appreciate that their main window into the microbial world was a bit foggy: laboratory culture, microscopic observations, and pathogenic actions each were strong selectors for specific microbial characteristics that they recognized as secondary qualities (to harken back to our discussion of Aristotles's natural kinds in chapter 2). In a phrase that captured the coming crises in microbial systematics, James Staley and Allan Konopka described The Great Plate Count Anomaly (12).

As early as the 1930s some microbiologists noted that their counts of bacteria made under the microscope did not reflect the number of bacterial colonies that they could grow on Petri plate cultures in the laboratory. These discrepancies were not trivial. . . often 100 to 1000 times different. Were there a lot of dead bacteria in these samples, bacterial carcasses that were no longer alive and capable of forming colonies? Simple chemical tests using certain dyes under the microscope indicated that these bacteria were still metabolically alive. It appeared, then, that laboratory culture methods were just missing many bacteria that did not oblige the scientists by growing on the culture media and under the conditions found in their laboratories (13).

Bacteriologists such as Staley and Konopka who were studying aquatic and soil bacteria were especially aware of such discrepancies and started to take the problem of The Great Plate Count Anomaly seriously. Archaea and various extremophiles provided examples of hard-to-cultivate microbes that hinted at the existence of even more problems. Soon bacteriologists began to speak of the "unculturables" (or at least the uncultured microorganisms). Clearly new approaches to find these previously unrecognized microbes were needed. Microscopic searches assisted by stains and antibodies as well as the higher magnifications provided by the electron microscope would yield new information that greatly increased the catalog of known organisms. Exploration of environments usually ignored by classical microbiologists such as various layers in the oceans, deep sea vents, hot springs, different soils, the human oral cavity and intestines, animal tissues such as the rumen microbes, and symbiotic relationships with other organisms, plant and animal. Many new bacteria were found in such neglected places. Still, the exact characterization and phylogenetic relationships of many of these new organisms were problematic. The answer to many of the problems of dealing with the unculturable microbes would be provided by Woese's new tool of genomic analysis of 16S rRNA. Even though an organism might not grow in the lab, if one could sequence its genes, it could be classified and fitted into the tree of life. Microbial diversity might still be comprehensible. This new approach, termed "metagenomics" will be considered in detail in the next chapter, but first we will turn to the problem of how many bacteria exist on Earth.

The question of how many bacteria exist on Earth is more than a trivial one. Bacteria comprise a significant mass of life on Earth, and as such, they consume and produce material that compete with, as well as supplement, other forms of life, including the human organism. Many soil bacteria, the "nitrogen fixers," capture nitrogen gas from the air and use it to make amino acids, the building blocks of all proteins. The blue-green bacteria in the oceans, the cyanobacteria, capture carbon dioxide from the atmosphere and convert it into sugars and then into other useful biological molecules. Some scientists have proposed that we control the growth of cyanobacteria in the oceans to modulate the carbon dioxide content of the atmosphere and hence reduce global climate change. In deep historical terms, these cyanobacteria are the source of much of the world's oil. Bacteria are also a source of genetic diversity and evolutionary "experimentation," producing new biological substances that may be useful or may be harmful; the larger the overall population, the greater the "opportunity space" for experimentation. The role of microbes in our world is certainly dependent on their number, locations, functions, as well as diversity, all factors that have attracted intense scientific interest since the microbial surprises of the 1960s and 1970s.

Since oceans cover much of the surface of the Earth, that might be a good starting place to begin counting the world's microbes. As a microbial culture vessel, the oceans are far from homogeneous, with vastly varying light, nutrient concentrations, temperatures, and pressures, but still we might get some estimate of the number of bacteria in the oceans by taking known averages for the number of bacteria that can be counted in diverse samples from different depths, different latitudes, and different climate zones. Multiplying a generally accepted average number of bacteria per milliliter of sea water (about a million per ml) by the total volume of the global oceans (about 1.5×10^{24} ml) gives the answer that there are about 1.5×10^{30} bacteria in the oceans of the Earth. Marine biologists estimate that about 98% of the weight of the biomass of the oceans is made up of microorganisms (bacteria, algae, and protozoa). We can get a more intuitive understanding of the enormous mass of bacteria in the oceans if we think of that mass in "Blue Whale Equivalents," that is, how many blue whales, the largest known animal species, would it take to equal the weight of all those microbes (14). Again, taking some average estimates for the weights of an average bacteria (4 picograms, or 4 one-thousand millionth of a gram) and the average weight of a mature blue whale (150 tons), there are about 4×10^{10} (forty billion) Blue Whale Equivalents of bacteria in the oceans; still almost inconceivable. In terms of numbers but not weight, the bacteria are outnumbered ten-to-one by viruses, mostly bacteriophages, viruses that prey on the bacteria, only about a million Blue Whale Equivalents of viruses (15).

Counting the bacteria on land is more problematic. The diversity of terrestrial ecology is vastly more complex than that of the oceans, so it is nearly impossible to get some single, reliable global estimate of microbial populations. We can, however, cite a few examples that give us a sense of the vast biomass of these unseen organisms all around us (or in us; consider our own digestive tract). It has been known since the early days of microbiology that the intestines of animals harbor bacteria that forage in the undigested residues of the food we eat, both producing more microbes and providing some byproducts that are essential for the host organism, for example the vitamin, biotin. Studies of human and animal feces reveal that fully half the wet weight of fecal material from normal, healthy individuals consists of viable bacteria that have grown in our gut every day.

Specialists who study soil microbes also find large and diverse populations of microbes in the soil, essential for the growth of plants. When we compost organic materials, we rely on the soil microbes to convert our "waste biomass" into more bacteria as well as byproducts that are useful to plants. Gardeners often say that feeding the soil microbes is the best way to think about how to fertilize the plants. Taken as a whole, soil microbes make up another large component of the Earth's microbial inhabitants.

HUMAN BEINGS AND OTHER ANIMALS

In the early years of microbiology, scientists were impressed by the specificity of various disease-causing microbes: humans suffered from cholera, but rabbits did not. Fowl plague affected birds but not people. Indeed, specific pathogenic patterns were taken as one of the essential characteristics used to identify and classify a microbe. Again, the decade of the 1970s was a time of reckoning for this viewpoint. More and more, microbiologists realized that this time-honored distinction of species-specific microbes was the exception rather than the rule. Humans were not so special after all; we share microbes with other animal species quite often, frequently to our peril. This realization came to be called the "One Health" concept. Epidemiologists as well as clinicians started to focus on animal epidemics and so-called "zoonotic diseases" to understand human diseases (16).

In the 1980s the "One Medicine" concept was first formulated in the veterinary medicine community, recognizing that human and veterinary medicine needed a combination of approaches to deal with zoonoses, diseases that can be transmitted from animals to humans, such as Ebola, Zika virus, COVID-19, and other "emerging" diseases. This movement was, to a significant extent, a result of the growing recognition of our ignorance of the diversity and extent of the microbial world that confronted science in the mid-twentieth century decades. Do we share our Earth with the microbes or do they share theirs with us?

NOTES AND REFERENCES

1. The term "gene" is used in several ways by the scientists who study heredity. This problem of "reference" was examined in detail by the noted philosopher of science Philip Kitcher; see **Kitcher P.** 1982. Genes. *Brit J Phil Sci* **33**:337–359. In the context here, the term gene most often refers to a stretch of a linear DNA molecule made up of linked subunits arranged in a specific sequence, reading like a sentence conveying instructions for cellular activities.

2. **Rokas A.** 2001. Getting it right for the wrong reason. *Trends Ecol Evol* **16**:668.

3. **Zuckerkandl E, Pauling L.** 1965. Molecules as documents of evolutionary history. *J Theor Biol* **8**:357–366.

4. **Sanger F, Brownlee GG, Barrell BG.** 1965. A two-dimensional fractionation procedure for radioactive nucleotides. *J Mol Biol* **13**:373–398.

5. **Fox GE, Pechman KR, Woese CR.** 1977. Comparative cataloging of 16S ribosomal ribonucleic acid: molecular approach to procaryotic systematics. *Int J Syst Evol Microbiol* **27**:44–57.

6. **Woese CR, Fox GE, Zablen L, Uchida T, Bonen L, Pechman K, Lewis BJ, Stahl D.** 1975. Conservation of primary structure in 16S ribosomal RNA. *Nature* **254**:83–86.

7. **Woese CR, Fox GE.** 1977. Phylogenetic structure of the prokaryotic domain: the primary kingdoms. *Proc Natl Acad Sci USA* **74**:5088–5090; see also **Woese CR, Magrum LJ, Fox GE.** 1978. Archaebacteria. *J Mol Evol* **11**:245–252.

8. See Reference 7.

9. In the PCR reaction, double strand DNA must be copied by a DNA polymerase, then new DNA must be heated to near boiling to separate the strands of the new DNA for entering the next cycle of the reactions. This process must be repeated 20-30 times to synthesize enough of the desired new DNA. Ordinary DNA polymerase is inactivated by the heating step so fresh enzyme must be added for each cycle. If the heat-resistant DNA polymerase from *Th. aquaticus* is used, such additions are unnecessary because the heat-resistant enzyme is still active for each subsequent heating cycle.

10. **Kouyama T, Kinosita K Jr, Ikegami A.** 1988. Structure and function of bacteriorhodopsin. *Adv Biophys* **24**:123–175; see also **Henderson R, Schertler GFX.** 1990. The structure of bacteriorhodopsin and its relevance to the visual opsins and other seven-helix G-protein coupled receptors. *Philos Trans R Soc Lond B Biol Sci* **326**:379–389.

11. **Postgate J.** 1993. *The Outer Reaches of Life.* Cambridge University Press, Cambridge, UK.

12. **Staley JT, Konopka A.** 1985. Measurement of *in situ* activities of nonphotosynthetic microorganisms in aquatic and terrestrial habitats. *Annu Rev Microbiol* **39**:321–346.

13. **Jannasch HW, Jones GE.** 1959. Bacterial populations in sea water as determined by different methods of enumeration. *Limnol Oceanogr* **4**:128–139.

14. **Abedon ST.** 2008. Phages, ecology, evolution, p 1–28. *In* **Abedon ST** (ed), *Bacteriophage Ecology.* Cambridge University Press, Cambridge, UK.

15. See Reference 14.

16. **Schwabe CW.** 1978. *Cattle, Priests, and Progress in Medicine.* University of Minnesota Press, Minneapolis, MN.

17. **Woese CR.** 1987. Bacterial evolution. *Microbiol Rev* **51**:221–271.

7 The New Microbe Hunters

Many of the microbial surprises of the 1960s and 1970s came as a result of rather traditional bacteriological practices: laboratory culture and isolation, microbial physiology, and exploration of pathogenic effects of the new isolates. Some discoveries, as we have already discussed, increasingly relied on methods and results from the newly developing field of microbial genetics. Genetics itself was being transformed in the 1970s from indirect study of the gene by studies of phenotypes to direct studies of the genotype (1).

This shift was made possible by the wholesale incorporation of biochemistry into the discipline of genetics. A direct attack on the chemistry of the gene and its two basic functions, heredity and directing the reactions of life, became possible. The basic chemical structure of the gene was elucidated in 1953 by the famous work on DNA by James Watson (1928–) and Francis Crick (1916–2004). The genetics of microbes rapidly refocused on DNA and its duplication to provide exact copies during cell division solved the Biblical conundrum of how like begets like: it is the result of the stereochemistry, that is, the chemical shape of the complementary building blocks, called nucleotides that pair up as part of DNA synthesis in every cell. Further, the biochemists soon provided an outline of how genes functioned to determine the machinery of the cell which carried out the reactions needed for life: energy metabolism, protein synthesis, and the like. The characteristics that microbiologists had been using for years to identify and classify bacteria were found to be encoded in the genes using a "genetic code" consisting of the linear order of the four nucleotide bases. In the late 1960s and early 1970s, biochemists

Magic Bullets, Miracle Drugs, and Microbiologists: A History of the Microbiome and Metagenomics,
First Edition. William C. Summers.
© 2024 American Society for Microbiology.

worked hard to devise methods to "read" the code directly from the DNA (and, as mentioned earlier, from RNA which was copied from DNA of the genes). DNA sequence analysis thereby bypassed the need for many of the approaches used by the classical microbiologists. In particular, the systematicists and evolutionists were able to get much closer to the biological facts on which they based their species classifications, their cladistic organizations, and their speculations about the evolutionary pathways in the ancestries of the organisms they studied. To understand the revolutionary nature of these methodological tools of microbial genetics, it will be helpful to review, briefly, the development of gene manipulation and sequencing that took place in the 1970s.

DNA CHEMISTRY AND GENETICS

As microbiologists came around to the notion that the identification, organization, and classification of microbes could be based on evolutionary relationships as was happening with more complex organisms, they faced a special problem: the minute size of their favorite organisms provided them with very few distinguishing characteristics that might be markers for distinguishing genes. Indeed, until mid-twentieth century, some biologists still doubted that bacteria had a reproductive system based on genes in the same way as higher organisms. But by then, the 1940s and 1950s, nutritional requirements, antibiotic responsiveness, and a few morphological markers in bacteria indicated that bacterial genetics was on its way forward. Rapid progress in this field was facilitated by the focus on some viruses that infect bacteria (2). The genes of these bacteriophages, together with those of all known forms of life, seemed to be made of the same chemical substances, the nucleic acids, DNA or RNA. The great impetus for more work on the chemistry and biology of nucleic acids came, of course, from the elucidation of the molecular structure of DNA by Watson and Crick in 1953. Genes conceived as units of function became identified as made up of chemicals, i.e., DNA, thus linking function to structure.

Instead of identifying and comparing microbes by their few visible characteristics together with their basic growth requirements, a very incomplete surrogate for a comparison of the genes themselves, microbiologists envisioned a classification system based on a direct chemical comparison of genes. The first hints that this approach would work were based on new methods of DNA chemistry worked out by physicists and chemists, not so much by biologists. The double-strand helix composed of two chains of DNA as proposed by Watson and Crick had a strict rule: the two strands bonded together in a perfect complementary way, like jigsaw puzzles. If one strand had the sequence of chemical subunits (the nucleotides that came in one of four different forms, designated A, G, C, and T for their trivial chemical names), say GAATC, the opposite strand would have the sequence CTTAG, specified by the rule that G and C have to pair, and A and T have to pair. One strand was a complementary copy of the other. This simple but beautiful chemically-required stereochemical principle is the mechanistic basis for all biological heredity, "how life begets life."

DNA HYBRIDIZATION AND GENE COMPARISONS

In the late 1950s Paul Doty (1920–2011) and his colleagues at Harvard were studying how to take the two intact strands apart and put them back together again. Heating DNA would break the bonds between the complementary strands and release the two strands separately from the rigid helix to allow them to be random coils, rather like limp spaghetti. They called this process the "helix-coil transition" or in the jargon of the time, "DNA melting." Soon they found that with intermediate temperatures and slow cooling, they could coax the strands back to the helical form of double-strand DNA, in perfect alignment of the original chemical sequence of nucleotide pairing. In analogy to very chemical processes of melting and cooling, they called this "annealing," or "renaturation" because heated biological molecules often became disordered and "denatured" so the reverse process might appropriately be called "renaturation."

In a rather forgotten paper in 1961 they reported a landmark result, important for the future of microbial genetics: if DNAs from two different species of bacteria were mixed, the renaturation did not go was well as the control mixture with only a single species (3). They could show that the DNA strands from different species renatured to form "hybrid" helices that were not as stable to heat as the non-hybrid, pure helices. In other words, they had an assay for DNA sequence mismatching, a direct comparison of the gene sequences between two species of bacteria. This method, called nucleic acid hybridization, which for a while was the only tool for direct comparison of gene sequences, was widely applied in other taxonomic studies, including birds and hominids (4). Still, it did not give direct gene sequence information, the key information that could allow gene by gene comparisons needed to distinguish apparently similar organisms and for accurate elucidation of evolutionary relationships. But the DNA chemists were hard at work and were also very lucky. By the time microbiologists woke up to their naiveté in the 1970s, methods were on the horizon that would pave their way forward.

RNA AND DNA SEQUENCE ANALYSIS

The way forward to gene sequencing would be made possible by three new technologies that soon appeared. A new class of enzymes was discovered that snipped DNA at specific sites to produce small fragments that were easy to sequence, such fragments could be linked to naturally occurring, self-replicating DNA molecules called plasmids and introduced into other bacteria for genetic study, and a biochemical technique was devised for essentially unlimited synthesis in the laboratory of any desired DNA molecule by a cyclic and repetitive process of *in vitro* DNA synthesis call the polymerase chain reaction, PCR. These three advances were so crucial that it is worthwhile to look at them in a bit more detail (Figure 7.1).

There are two challenges at the start of any attempt to determine the sequence of the nucleotide building block units in DNA or RNA: obtaining a homogeneous sample and finding a manageable size molecule to analyze. Since both the RNA and DNA molecules in genes are long, linear chains of the four kinds of subunits, very long molecules of highly repetitive units can get confusing if the chains are too long...like trying to read and copy a string of digits without making a mistake. Easy for something like a phone number but hard for something like the first 200 digits of the expansion of the number called pi (π). Because most

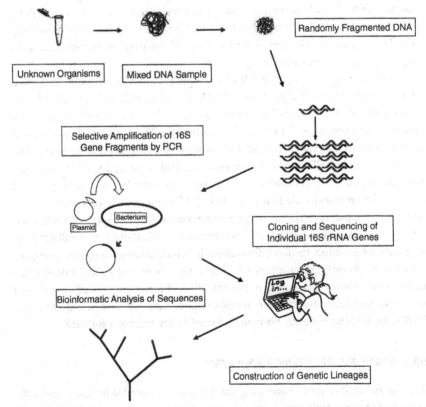

FIGURE 7.1 *Steps in metagenomic analysis from original sample to production of evolutionary diagram.*

chemical methods require not a single molecule but a large number of molecules to yield a detectable answer, it is important to have all those molecules be identical rather than a mixture of sequences. Although nature has given scientists a few examples of discrete small DNA and RNA molecules, some of which can be isolated in relatively large, homogenous masses from cells (for example the RNA molecules that carry amino acids in protein synthesis called transfer RNAs or tRNAs), most interesting nucleic acids need to be broken down into reproducible chunks before their sequences can be figured out.

These two roadblocks, the size and the homogeneity problems, started to yield to biochemists in the 1960s. A few enzymes from the digestive systems of animals or fungi were discovered that could be used to cut the RNAs at specific sites, resulting in reproducible fragments of sequencing size. Further, separation methods, perfected during the World War II work on uranium chemistry for atomic bombs, were applied that could allow large scale purification of these fragments. Short sequences of RNA began to be determined and eventually assembled into longer, sometimes complete sequences of rRNA and tRNA molecules (5). With hindsight, these methods look unbelievably primitive and cumbersome, but at the time were both exciting and promising of things to come.

While RNA sequences were useful, the real prize was the DNA sequence. DNA, however was problematic since no digestive enzymes had been found with the specificity to cut DNA at specific sequence sites, and DNA was incomparably long compared to the full-length RNA starting molecules. Even if enzymes could be found that cut DNA at specific sites, there would be too many different fragments to yield to separation into homogenous samples. Again, nature was kind, and human ingenuity came to the rescue. This phase of the story illustrates the importance of fundamental curiosity-driven science to produce totally unexpected practical results.

ENZYMES THAT CUT DNA PRECISELY

Starting in the early 1950s a few microbiologists who studied bacteriophages, the viruses that infect bacteria, (a small community of scientists at the time) noted an odd phenomenon in which the infectivity of the bacteriophage for fresh bacterial host cells depended on the particular strain of bacteria in which the bacteriophage had just previously infected and replicated. The few bacteriophages that did manage to infect the new bacteria, emerged fully infectious for the same strain of bacteria. These phage biologists reasoned that the new bacteria "restricted" the growth of the phage grown on the earlier host, and that each host was able to put its own "modification" or some sort of identification on virus that had grown in it. Modified phage could escape being restricted by the bacteria that normally restricted unmodified phage. Most microbiologists yawned. Phage were weird anyway, and this kind of odd infectivity pattern was just one more manifestation of that weirdness. A small group of enthusiasts persisted. They found, for example, that the restriction was the result of the immediate breakdown of the incoming viral DNA as it entered the restricting bacterium. Two questions arose: what is the enzyme machinery that degrades this DNA so specifically? And how is the modified, i.e., resistant, DNA marked to avoid degradation? The answer to these two questions would soon revolutionize modern biology, lead to several Nobel Prizes, spawn a multi-billion-dollar global biotechnology industry, and lead to a new era in medicine. No more yawns.

It turns out that many bacteria have collections of DNA-degrading enzymes that are exquisitely designed to attack and cut up invading DNA but not attack the bacteria's own DNA. It does so by marking (modifying) its own DNA with addition of chemical groups (usually methyl, CH_3, groups) to the same specific sequence stretches the degradation enzyme attacks thereby protecting it from cutting. In order for these defensive enzymes to distinguish friend from foe, to allow gene exchanges with close relatives, but not with distant and incompatible forms, these enzymes are exquisitely specific, able to recognize specific stretches of DNA usually 4 to 8 nucleotide bases long. On a simple statistical basis, these stretches should occur in any random DNA molecule about once in every few hundred to once in every few thousand bases. These enzymes, dubbed restriction endonucleases, or simply restriction enzymes, do just that: they cut DNA at specific gene sequences and produce defined pieces of DNA of precise lengths and sequences. This process would be as if one searched all the volumes of Shakespeare's work in the library and snipped at the start of every phrase "to be." One result of this desecration of the Bard's work would be a pile of small bits of texts "to be or not," easy to identify by size, sweep up, and read.

PLASMIDS: CARRIERS OF NEW GENES

The discovery of these restriction enzymes, in addition to explaining an interesting fact about microbial genetic policing, provided molecular geneticists with a powerful tool for breaking the huge genome into manageable pieces to sequence with their new methods. Once again, the amazing diversity and evolutionary ingenuity of microbes surprised even the diehard optimists.

Another nearly simultaneous stroke of luck provided one more tool needed to advance the revolution in genetics that was underway. Since at least the 1940s, it was known that many cells, including microbes had small, perhaps auxiliary DNA molecules, separated from the main chromosomal DNA molecules. The functions and structures of these DNAs were disputed and certainly poorly understood.

In a major clarifying summary, Joshua Lederberg, in 1952, proposed that the term plasmid "as a generic term for any extrachromosomal hereditary determinant" (6). By the 1960s, some of these plasmids were identified in bacteria as the genetic carriers of the information providing antibiotic resistance to that bacterium such as in the case of the R-factors we met previously. Even more significant was the fact that these plasmids could move from cell to cell without involving the main hereditary apparatus of the cell, what became known as "lateral" or "horizontal" gene transfer (in distinction from "vertical" transmission of parent to offspring). Antibiotic resistance did not only arise by random mutation, but could also spread like an infection among neighboring bacteria, clearly an important medical and epidemiological concern. Aside from its medical importance, lateral gene transfer by plasmids was soon recognized as a possible way to think about introducing laboratory-constructed chunks of DNA into specific bacteria.

The conjunction of restriction enzyme cutting of DNA into specific pieces and the capacity of plasmid DNAs to move between bacteria, soon led to the first major step in genetic engineering since artificial, selective breeding was developed millennia ago by our agricultural ancestors. Specific gene sequences could be snipped from one cell, joined (using a newly discovered linking enzyme called DNA ligase) to a plasmid DNA to make a hybrid, or "recombinant" DNA, molecule that could be introduced into another cell where the augmented plasmid would take up residence, having carried a foreign gene sequence into a new cellular environment. The foreign gene sequence, that is its DNA, or often the gene product, a new protein, could be mass-produced by growing quantities of the newly modified cell. This process came to be known as "molecular cloning" because a single molecule of DNA could be introduced into a single cell which, when isolated and grown up in laboratory cultures could produce huge quantities of genetically identical plasmid DNA molecules, all derived from the one "founder" recombinant molecule.

AMPLIFYING GENES *IN VITRO*

Molecular cloning made it possible to study rare and even unculturable bacteria by simply isolating a few (or even a single) DNA molecules from a tiny sample, moving DNA fragments cut out by restriction enzymes into an easily grown host bacteria, often the geneticists' favorite, *Escherichia coli*, where unlimited amounts of the DNA could be sequenced for study and the gene products, the proteins, could be made, purified, and studied. Gene sequence data and microbial descriptions and identification of genetic relationships rapidly accumulated in the 1980s. The third of the essential tools of the new genetics was devised in 1983 when Kary Mullis (1944–2019) showed that single molecules of DNA can be repetitively copied in the test tube using known DNA-synthesizing enzymes (polymerase) and appropriate reaction conditions (7). Because each repetition of the synthetic reaction doubles the amount of the starting DNA, copying the exact sequence each time, the amount of input sequence grows exponentially with each repetition (2^n, where n is the number of repetitions, or reaction cycles). This tool, the polymerase chain reaction (PCR), allowed scientists to bypass the laborious molecular cloning step in many experiments. It had the additional advantage of being automated to handle many samples simultaneously, a so-called "high-throughput process."

With restriction enzymes to cut DNA at specific sites, molecular cloning to isolate and study almost any gene of interest in a standard laboratory context, and PCR to amplify specific DNA sequences on a massive scale, microbial genetics underwent a major revolution in the decade of the 1980s. No longer was microbiology dominated by culture media and growth requirements, microscopic morphology and staining properties, instead it became nearly a subdiscipline of biochemistry, and for some, of computer science. The hunt for new microbes was not driven by the emergence of new diseases as in the 1960s, but the focus shifted to understanding the previously unrecognized, even unimagined, microbial world around us. As we will see, this new focus would, however, directly inform understanding and responses to later waves of emerging infectious diseases.

DNA and RNA sequence analysis developed in less than two decades from an esoteric skill available in only a handful of laboratories across the world, to a routine tool, automated by sophisticated machines that relied on both the early methods of sequence analysis and then entirely new "next generation" sequencing approaches (8). Gene sequencing became outsourced as a service function rather than a research activity in itself. It has now gone from the lab bench to the bedside, as the saying goes. Just as growth properties, morphology, and staining were routine to a previous generation of microbiologists, today, nucleic acid sequencing is an indispensable tool for microbiologists in the twenty-first century.

BIOINFORMATICS AND METAGENOMICS

Even with DNA sequences of several hundred bases in length, it was difficult to keep records, copy the sequences and edit them manually without errors. The "mutation rate" for human copying of DNA sequences far exceeds that of the cellular gene replication processes. Even in the earliest days of both protein and nucleic acid sequence studies, record-keeping, comparisons, and analysis, called for some level of computational help. Initially, simple file storage and editing tools were sufficient, but soon the handling and manipulating of the sequence generated in multiple experiments showed the need for programs specifically designed for these tasks. Some laboratories saw the future need for even more advanced computing and began to employ computer scientists as integral collaborators. Gradually, these specialists developed an identity for their field as bioinformatics. The history of bioinformatics is a subject in itself, but it is important when discussing modern molecular and microbial genetics to recognize the essential role of bioinformatics in microbiology. The computational problems are significant when manipulating databases of with complex comparisons of millions to billions of sequences needed to align the sequences generated in a whole-genome sequencing project. New methods of massively-parallel sequence analysis needed for study of microbial population structures and dynamics generate gigabytes of data that much be handled in various ways and pose both data storage and analysis problems. When we look at summary tables or graphs of microbial diversity data, it is well to keep in mind that massive computational work went into making comprehensible the masses of sequence data behind them.

As gene sequencing has advanced through mechanization and automation, and as massive amounts of data have been collected, microbiologists no longer need to focus their attention on a single species, or small group of microbes. Instead, the massive data can be "mined" for relationships, correlations, oddities, and functional properties that might emerge from considering all the data simultaneously (9). Such analyses, of course, require computational and data-handling capabilities of computers and computer scientists. The approach to microbial identification and organization based on massive genomic data sets is known as metagenomics. Because this approach uses DNA sequences as the input data, metagenomics enables study of uncultured microbes, populations of microbes from specific ecological niches, and construction of subtle evolutionary relationships. Metagenomics first appeared in the 1990s when sequence data from the growing list of completely sequenced microbes became too complex to analyze "by hand," that is, by a single human, with paper, pencil, and an imagination. Enter the bioinformatician. Simple computer programs to copy and search for sequences, often designed by enthusiastic, self-taught microbiologists, gave way to sophisticated programming efforts by professional computer scientists who provided programs that could align sequenced fragments into complete genomes, find sequences that are likely to be protein-encoding genes, arrange sequences in likely evolutionary schemes, and the like. Computing power is now a requisite tool of many microbiological research teams.

Metagenomic challenges include such key questions as i) how large should a data sample be to provide secure conclusions? ii) should there be some sort of filtering of the data to remove or simplify known or suspected categories of data prior to a specific analysis? iii) what kinds of "data about the data" called metadata are needed? iv) how can the quality of the data be estimated, that is, do the data accurately sample the population of interest? The bioinformatics algorithms to identify gene-coding sequences (often called "open reading frames"), to determine contiguous sequences to be used to assemble more complete genomes from sequenced fragments, and to measure diversity of a given sample population, all require starting assumptions that are not always agreed upon by either the biologists or the computer scientists (10).

As the massively parallel, automated DNA sequence determinations become routine, inexpensive, and medically useful, metagenomic information and approaches will become commonplace in public health, personalized medicine, and environmental studies. Now one can truly speak of a microbiome not just as a concept, but in specific terms, with precise characterizations of microbial species, their frequencies of representation, and the changes and pathologies underlying such data. Bioinformatics still struggles to devise ways to describe and graphically display these data in easy-to-comprehend fashions. Both the analytical algorithms and the data display approaches reflect the uses intended for the metagenomic data. Evolutionary biologists need their information to show phylogenetic linkages while clinicians will need data from a patient's body to guide diagnosis and potential therapies.

One example of a widely used tool for metagenomics is a large data base associated analytical software called KBase, described as "a collaborative, open environment for systems biology of plants, microbes and their communities," is maintained and supported by the U.S. Department of Defense. Interestingly, the majority of the scientists involved in this project are associated with just three research programs: the Joint Genome Institute in Berkeley, California, the Argonne National Laboratory in Illinois, and the Lawrence Berkeley Laboratory, also California. The latter two laboratories are sites of "big science" long associated with high-energy atomic research, programs that were pioneers in the use of computers to handle big data projects. A recent (2020) report from this consortium, modestly titled "A Genomic Catalog of the Earth's Microbiomes" gives an example of the success, scope, and impact of the metagenomic approach to microbiology (11).

With KBase and other systems, this group of 32 authors describe the reconstruction of over 10,000 bacterial and archaeal metagenomes by so-called shotgun sequencing (12). These samples for metagenomic analysis came from diverse habitats, environments, and agricultural soils to allow a relatively unbiased picture of genomes representing 12,556 novel candidate species-level organisms. This collection spans 135 different phyla, and expands known phylogenetic diversity of bacteria and archaea by 44%. They have grouped their results in various ways, but one basic grouping is based on the ecological source of the genome, defining a biome as a potentially interacting community of nearby organisms. They use four broad categories: aquatic; engineered (that is biomes in man-made communities); host-associated; and terrestrial. Aquatic biomes include subgroups for freshwater and marine genomes; engineered biomes include, among others, wastewater and lab enrichment samples; host-associated groups include plants and humans; and the terrestrial category of biomes has caves, plant litter, and soil as well as several others.

In the next chapter we will take a closer look at some of the recent discoveries made possible by these new microbiological techniques to better understand the world of microbes that surround and affect everything we do.

NOTES AND REFERENCES

1. **Streelman JT, Kocher TD**. 2000. From phenotype to genotype. *Evol Dev* **2**:166–173.
2. **Summers WC**. 2023. *The American Phage Group: Founders of Molecular Biology*. Yale Univ. Press, New Haven, Conn.
3. **Schildkraut CL, Marmur J, Doty P**. 1961. The formation of hybrid DNA molecules and their use in studies of DNA homologies. *J Mol Biol* **3**:595–617.
4. **Sibley CG, Ahlquist JE**. 1984. The phylogeny of the hominoid primates, as indicated by DNA-DNA hybridization. *J Mol Evol* **20**:2–5; see also **Sibley CG, Ahlquist JE, Monroe Jr. BL**. 1988. A classification of the living birds of the world based on DNA-DNA hybridization studies. *Auk* **105**:409–423.
5. **Holley RW, Apgar J, Everett GA, Madison JT, Marquisee M, Merrill SH, Penswick JR, Zamir A**. 1965. Structure of a ribonucleic acid. *Science* **147**:1462–1465.
6. **Lederberg J**. 1952. Cell genetics and hereditary symbiosis. *Physiol Rev* **32**:403–430.
7. **Mullis K, Faloona F, Scharf S, Saiki R, Horn G, Erlich H**. 1986. Specific enzymatic amplification of DNA in vitro: the polymerase chain reaction. *Cold Spring Harb Symp Quant Biol* **51**:263–273.
8. **Hutchison CA III**. 2007. DNA sequencing: bench to bedside and beyond. *Nucleic Acids Res* **35**:6227–6237.
9. Data mining is a phrase that captures a recent experimental approach that is often placed in opposition of "hypothesis driven" experimentation. Data mining is characterized by, supposedly, collecting masses of data without a guiding hypothesis followed by looking for "interesting" characteristics in the data set. Often computer searches for regularities or other characteristics are involved. The absence of a "hypothesis" in data mining is clearly untrue; hypotheses enter many aspects of data mining, from the first step in selecting the data to be collected to the final examination of the data based on notions of what counts as relevant or interesting.
10. **Wooley JC, Godzik A, Friedberg I**. 2010. A primer on metagenomics. *PLOS Comp Biol* **6**:e1000667; see also **Simon HY, Siddle KJ, Park DJ, Sabeti PC**. 2019. Benchmarking metagenomics tools for taxonomic classification. *Cell* **178**:779–794; and **Fleischmann RD, et al.** 1995. Whole-genome random sequencing and assembly of *Haemophilus influenzae Rd. Science* **269**:496–512; and **Chivian D, Jungbluth SP, Dehal PS, Wood-Charlson EM, Canon RS, Allen BH, Clark MM, Gu T, Land ML, Price GA, Riehl WJ, Sneddon MW, Sutormin R, Zhang Q, Cottingham RW, Henry CS, Arkin AP**. 2023. Metagenome-assembled genome extraction and analysis from microbiomes using KBase. *Nature Protocols* **18**:208–238.
11. **Nayfach S, et al.** 2020. *A Genomic Catalogue of Earth's Microbiomes–Introductory KBase Narrative*. US Department of Energy Joint Genome Institute (JGI) US Department of Energy Systems Biology Knowledgebase. http://dx.doi.org/10.25982/53247.64/1670777
12. **Quince C, Walker AW, Simpson JT, Loman NJ, Segata N**. 2017. Shotgun metagenomics, from sampling to analysis. *Nat Biotechnol* **35**:833–844.

8 Global Microbiology

Microbiologists have long considered their favorite forms of life to be of global interest and importance. The new science of metagenomics has amply confirmed that belief. In 1998 William B. Whitman, David C. Coleman, and William J. Wiebe published a "Perspective" paper in the prestigious *Proceedings of the National Academy of Science* which pointed to the magnitude of microbial life on Earth, referring to the "unseen majority" (1). They collected information from studies of microbes in various aquatic environments, in the soil, and in subsurface habitats and, with some assumptions to bridge the gaps in data, arrived at some approximations that were both startling and interesting: there are roughly 5,000,000,000,000,000,000,000, 000,000,000 (that's 30 zeros) single-cell organisms on Earth, containing carbon, the main element in living things, weighing about 4.5 billion (giga) tons in total. These scientists pointed out several consequences that follow from these numbers. First, the enormous number of individual cells that multiply as single genetic units means that the chances of a given genetic variation somewhere on Earth is high. A mutation that is rare in the tiny populations of microbes that can be studied in the lab has a much better chance of happening in nature. Second, this capacity of microbes to sustain and propagate genetic variation may impact models used in phylogenetic analyses. Third, because microbes are important in many chemical transformations in the environment including both biological and geological reactions, the abundance and distribution of these microbes are important facts for humans to appreciate as we continue life on Earth.

Magic Bullets, Miracle Drugs, and Microbiologists: A History of the Microbiome and Metagenomics,
First Edition. William C. Summers.
© 2024 American Society for Microbiology.

In the twenty years following this seminal paper, scientists from various disciplines have developed information on the microbial populations in more and more diverse and representative environments. In 2018, the question of the global microbial population was again revised. Scientists have come to realize that "a quantitative description of the distribution of biomass is essential for taking stock of biosequestered carbon and modeling global biogeochemical cycles as well as for understanding the historical effects and future impacts of human activities" (2). From such recent studies it has become clear that microbes comprise a surprisingly significant fraction of Earth's biomass. Further, the estimates of microbial carbon from two decades earlier are certainly about ten times too low. Although all such estimates are fraught with uncertainty, it is possible to make some reasonable estimates, and one recent authoritative summary calculated that the inhabitants of the microbial world (bacteria and archaea) contain about 77 billion (giga) tons of carbon and comprise about 17% of Earth's biomass (as measured by weight of carbon) exceeded only by plants which make up 83%. Animals make up a paltry 0.3%, represented almost entirely by arthropods (insects, spiders, and the like) (3) (Figure 8.1).

We humans rarely encounter such masses of microbes, but they are important none the less. The great bulk of Earth's microbes, it turns out, are in the deep sub-surface, underneath the oceans and in the massive underground water reservoirs

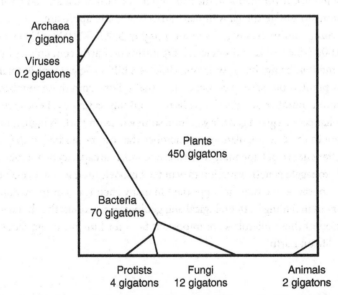

FIGURE 8.1 *Graphical representation of global distribution of biomass by different kinds of organisms. Weights are proportional to the areas on the Voronoi diagram. Redrawn from Figure 1 in reference (2), with permission © 2018 Bar-On, Phillips, and Milo.*

called aquifers. From the limited studies done on such biomes, the microbes in these environments are surprisingly slow growing. While lab bacteria grow at the doubling rate of minutes to days, these subterranean organisms usually turn over at rates of once in a few months to once in a thousand years. These seem to be just barely existing, perhaps carrying out basic metabolism, or even in dormant states called spores. Some of these organisms, having captured carbon from their inorganic environment and converted it to the organic molecules of life, eventually die, and under the heat and pressure of the depths where they have lived, give rise to deep deposits of petroleum, later to be pumped up by humans to burn for energy.

Scientists now use the term "biome" in reference to the collection of life forms that exist together in a particular "community." Earlier terms such as "ecological niche" and "local environment" were equally imprecise but intended to describe anti-reductionist thinking about interacting, diverse populations, living under external constraints and contexts. Microbial life in soil has been called the rhizobiome, for example, and that in the gut of mammals, the gut biome. As the new science of metagenomics is providing the tools to study these complex, diverse but interacting microbial populations, focused analyses are yielding interesting and potentially important advances. Figure 8.2 shows in a simplified way the prevalence data for various microbial species in a specific sample. A brief survey of three different microbiomes will help illuminate some of this recent work.

TALES OF THREE BIOMES

Starting in the summer of 1930 C.B. van Niel (1897–1985) a young Dutch microbiologist at Stanford University led a small group of students into the Pacific Ocean at the Hopkins Marine Station, a research laboratory on Monterey Bay in California, where each student would scoop up a pail of sea water, sand, and sediment to take back to the lab for microbial analysis as part of van Niel's famous course in general microbiology (Barbara J. Bachman, personal communication) (4). Generations of students in this course came to appreciate the diversity of the microbial world and the multiple interactions displayed by new and unknown organisms encountered in a simple bucket of stuff from the sea. Many of the ideas and conceptual frameworks of twenty-first century metagenomics and the biology of the microbiome can be traced to van Niel's influence and emphasis on global microbiology as a science (5).

Oceans cover about 70% of Earth's surface and contain about 97% of Earth's water, so it is not surprising that the microbial inhabitants of oceans are important to nearly all things global, both living and nonliving. Oceans, of course, have been seen as containing vital microscopic life for a long time: the component of marine life called plankton is well-known as a concept but rather vaguely understood in

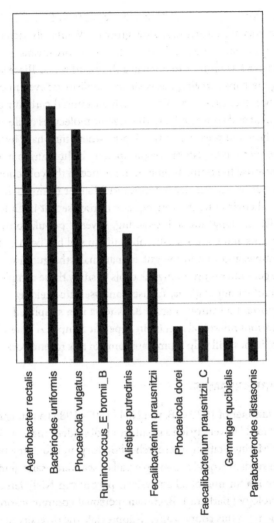

FIGURE 8.2 *Metagenomic data display (simplified) of the ten most prevalent procaryotic species in a human gut microbiome. Drawn with data from (40).*

many of its details. Plankton as a category of organisms is not defined by biological properties or phylogenetic descriptions, but simply as unattached and nonmotile organisms that simply drift in the water. A great proportion of these organisms are, indeed, the same ones that microbiologists have traditionally claimed for their discipline: bacteria, archaea, algae, and protozoa. These organisms of the marine biome have at least two important functions: they are food at the bottom of "the food chain" supporting other organisms that are important human and animal food sources (i.e., fisheries) and they are important ecological components in complicated bionetworks that process and regulate the flows of organic molecules, utilization of solar energy by photosynthetic organisms, and by this and other mechanisms contribute to the atmospheric carbon dioxide blanket that controls global temperatures. The new tools of metagenomics provides ways to probe more deeply into the complex interrelationships of the marine microbiome providing new insights in to the vast resources of Earth's oceans (6).

The oceans are not homogeneous, and their physical and biological properties vary with latitude, depth, season, geology, surface runoff, and other local influences. One crucial factor in the web of life in the oceans is the availability of sunlight that is essential for photosynthesis and the conversion of carbon dioxide into carbohydrates, the organic compounds that feed the entire food chain of the seas. Sunlight penetrates only a relatively thin layer of surface water, depending mostly on the turbidity caused by suspended material both organic and inorganic. This transparent zone at the surface has been called the euphotic zone (from the Greek, well-lit), a layer that usually ranges from about 150 to 650 feet deep. It is in this layer of the ocean that the green plants of the seas, the algae, live, absorbing the energy from the sun, and making the building blocks for most of the rest of ocean life. Very important, too, is the fact that these organisms trap atmospheric carbon dioxide for this process; they are a major factor in controlling the global carbon dioxide balance and as a result, the greenhouse layer regulating global temperatures.

Most bacteria and archaea in the euphotic zone rely on the photosynthetic organisms to produce certain organic nutrients that these bacteria and archaea cannot make for themselves. This results in a network of interdependent populations of diverse organisms on which much of global life depends. The marine microbiome contains species that degrade an astounding array of substances from large dead marine animals, such as whales, to microplastics, cardboard trash, and discarded antibiotic medications. Often marine biologists speak of "detritus" to capture just how varied and hard-to-describe the materials these microbes encounter can be. The diversity of the oceanic microbiomes, increasingly revealed by metagenomic studies, shows that there are complex flows of substances in and out of identifiable communities of microbes. An essential nutrient, iron, for example, is preferentially taken up and stored by the algae known as diatoms. They are particularly good at trapping and hoarding iron, to the extent that "blooms" of diatoms can occur when the competition for iron becomes acute and other organisms die off, leaving the existing iron for the diatoms.

Ocean microbes do not just inhabit the surface where the sun shines. With time, many microbes (as well as other particulate objects) sink to the ocean floor, entering an entirely different environment. They take with them, of course, materials from the surface layers of the ocean, and deposit it at the bottom of the sea. Much of this material is in the form of organic molecules from decomposing bacteria and archaea which were assembled in the euphotic zone from carbon atoms originally sequestered from the atmosphere. Much of the world's oil, for example, was formed over the millennia from the accumulation of marine bacteria which settled to the bottom of the oceans and were converted, over time and under the great pressures of those depths, into petroleum. The growth dynamics and species composition of the marine microbiome, we now know, has an important role in both the biological components of the oceans but also in some of the chemical balances there. When microbes take up and use various elements in their metabolism, carbon, nitrogen, sulfur, phosphorus being some of the major elements taken up in bulk, they may sequester these substances that are dissolved in ocean water, and if the microbes then fall to the bottom, can make the shallower zones nutrient-poor, starving other forms of life. Conversely, some microbes may sequester toxic materials that are safer for the planet to have locked away on the ocean floor, taken there by microbial remediation.

There is no better example of the complexity as well as the ecological importance of marine microbiomes than the recent work on coral reefs and their global status. Corals are tiny marine invertebrate animals that we mostly recognize by the complex and sometimes massive hard skeletons in the form of reefs that they construct from secreted calcium carbonate. The animal itself is called a polyp, a tiny sac-like organism only a few millimeters in diameter and a centimeter or so in height. Each polyp secretes an exoskeleton near the base, where it is attached to some fixed object. Over generations, communities of polyps build massive, hard structures. Corals can reproduce both asexually by fission as well as sexually by releasing sperm and eggs into the water which fuse to form zygotes that ultimately float away to form new colonies of mature polyps. Since corals are mostly sessile, immobile organisms, they depend on foraging for nutrients from their immediate environment. In the words of two coral scientists, "Corals are fundamental ecosystem engineers, responsible for constructing intricate reefs that support diverse and abundant marine life" (7). Coral reefs are sites of a dynamic collection of symbiotic organisms including bacteria, viruses, and eucaryotic protists that evolutionary biologist Lynn Margulis (1938–2011) termed the coral holobiont (8). This rich microbial community provides nutrition for the coral but in turn the coral reef structure provides support for the coral marine microbiome. One study estimated that the coral microbiome contains 25% of all the known marine species (9). The microbiome of the coral reef is now considered to be crucial to the survival and resilience of these reefs to climate change, pollution, and ocean acidification. Since coral polyps reproduce relatively slowly, of the order of 4–6 years for sexual reproduction, the capacity for genetic adaptation to changing conditions is likely too slow to respond meaningfully to external challenges. When one considers the entire ecosystem, the coral holobiont, the microbiome provides more adaptive and rapid responses. Microbes, because of their numbers and rapid reproduction rates, are able to generate genetic diversity to meet new challenges on a time scale that can effectively mitigate many challenges to the holobiont. As explained by Nicole Webster and Thorsten Reusch in a recent review of reef biology, "The functional basis of most reef bacterial symbioses centres on cycling of essential nutrients such as carbon, nitrogen, sulfur, and phosphate in addition to passage of trace metal, vitamin synthesis, provision of other cofactors and production of secondary metabolites" (10). They emphasize that by consideration of the coral holobiont rather than the isolated species, the dynamics of the microbiome can help explain and understand the processes that promote and limit coral adaptations. A diverse microbiome likely contains representatives of species that can assist in adaptation by simply increasing its fraction of the microbial population, a phenomenon they call "symbiont shuffling." If entirely new microbial species invade the microbiome and prove useful under new challenges, the microbiome will undergo "symbiont switching" during the process of acclimatization. Bacteria, they point out, have another option beyond simple population shifts: rapid genetic changes. Since bacteria usually reproduce by asexual means, simple cell division, they usually have only one copy of a given functional gene, that is, they are "haploid" so any functional change in that gene, for better or worse, is immediately seen in the next generation of offspring. In addition to this rapid evolutionary lifestyle, many bacteria are notoriously promiscuous, allowing foreign DNA to enter their gene pool by horizontal (or lateral) gene transfers, processes sometimes mediated by viruses and sometimes by rather rudimentary direct transfer of self-replicating DNA molecules that act as auxiliary mini-chromosomes (plasmids). Interestingly, from the holobiont point of view, the community can rapidly adapt to a new more biologically resilient state, and because the entire microbiome structure has evolved to a new ecologically stable state, it can propagate itself enabling "transgenerational acclimatization." As these authors point out, this way of responding to environmental challenges "does not required co-evolution of the entire holobiont as an individual unit. Rather specific microorganisms or microbial genes that provide a fitness advantage to the host under the conditions of elevated sea temperatures and ocean acidification could be maintained and passed to subsequent generations, ultimately resulting in transgenerational acclimatisation of the host" (11).

ROLE OF BACTERIOPHAGES

Thus far, we have focused our attention on the bacteria and archaea in aqueous microbiomes, but many studies have found that viruses that prey on these bacteria and archaea are at least ten times more abundant than their hosts. Numerically, these viruses are probably the most abundant organisms on Earth (12). Recent studies show that these procaryotic viruses, the bacteriophages, play crucial roles in the community dynamics of these aqueous microbiomes (13). These viruses, phages for short, most often attack bacteria and archaea, reproduce inside these microbes and then promptly disrupt, kill, and lyse the host cells. This lysis or dissolution has a key role to play in recycling organic compounds, releasing them from the carcasses of the killed microbes back into the water for use by other organisms. The fact that phage are often highly species-specific in their choice of hosts, the dynamics of the microbiome diversity can be driven by the population of phages that prey upon it. Further, bacteria that eventually sink to the ocean floor represent a removal of organic nutrients from the euphotic regions where they were synthesized. The phage-mediated lysis seems to control the proportion of microbes that sediment and hence modulate the flow of nutrients from the surface zones to the deep ocean floor. Phage lysis of these microbes allows recycling of nutrients in the more bioactive surface regions. Some scientists estimate that between twenty and fifty percent of some species of surface microbes are recycled by phage lysis every day (14).

While the oceans contain most of the water on planet Earth, the fresh water supplies and their suitability for biological support is of crucial import as well. Maintaining suitable water for drinking and safe disposal of contaminated wastewater have been a major concern of microbiologists almost from the beginning of the discipline in the latter part of the nineteenth century. Under the rubric of "sanitation" microbiologists have studied water purity and sewage disposal as two central topics of interest.

As soon as microbes were discovered as agents of disease, their presence in fresh water was of concern. How many bacteria and of what species characterize acceptable drinking water? A leading English pathologist, German Sims Woodhead (1855–1921), an authoritative early microbiologist, wrote in 1891 "It is sometimes said that if water does not contain more than one thousand organisms per cc. that it may be used with safety for drinking purposes, whilst it must be borne in mind that this thousand organisms may contain a larger number of pathogenic organisms, while on the other hand five thousand organisms in this same quantity might not include a single pathogenic germ" (15). Woodhead recognized that microbial diversity was a key factor in understanding water microbiology. He cited another leading microbiologist of that era, Walter Migula (1863–1938) who studied microbial diversity explicitly: "After examining 400 springs, wells, and streams, W. Migula concluded that when there are more than ten species of bacteria in any sample of water, especially when these species are not ordinarily met with, the water should not be used for drinking purposes. [...] He found in all 28 species, and observed that the number of colonies does not by any means correspond to the number of species, though in some cases it undoubtably does" (16).

Water purity, even now, is characterized by its microbiological properties. Bacterial counts and species diversity continue to be used to test drinking water for human safety and palatability. Bactericidal agents such as chlorine compounds are frequently added to municipal water supplies to inactivate potentially harmful microbes. At the site of use, heat, filtration, and solar exposure are common anti-microbial methods for water treatment. Microbe-free water is a goal of modern hygiene in the mind of many. As metagenomic analysis is beginning to be applied to drinking water, the advantages of this new method is becoming clear: the diversity of microbial contamination in various types of waters is being discovered, along with the recognition of a myriad of unculturable microbes that may be troublesome. Water pipe contamination, especially from biofilms that resist chlorination efforts are likely to be more detectable by metagenomic methods. Additionally, viruses, fungi, and protists can be identified without the need for elaborate culture methods.

One recent study provides examples of such a comparative analysis of different water samples: tap water, drinking fountain water, sparkling natural mineral water, and nonmineral bottled water were compared. Each source showed a distinctly different microbial profile. Interestingly, none of the samples showed evidence of fecal contamination, which would be indicated by presence of *Escherichia coli* or other enterococci species. Reassuringly, too, genes for antibiotic resistance were not detected in any of the samples. Only the sparkling natural mineral water con-tained any detectable archaea species (17). Another study of an immediate sub-surface aquifer along the Colorado River, likely a source of well water for local communities, showed a large diversity of microbial genomes by metagenomic analysis. Over 2,500 nearly complete genomic sequences, representing the major-ity of known bacterial phyla as well as 47 newly discovered phyla-level lineages were found in subsurface samples. The authors of this study note "The tremen-dous novelty of microorganisms observed in the aquifer ecosystem highlights the potential for biological discovery in the terrestrial subsurface. Given the novel phy-logenetic diversity of the studied organisms, the genomes reported here represent a vast treasure-trove that could be mined for biotechnological applications and for potential strategies for genome-enabled cultivation of novel organisms" (18).

While the presence of microbes in drinking water may be a problem, the need for microbes for biological processing of wastewater, i.e., sewage, is also an impor-tant consideration. As we have seen, microbial interconversions of organic matter in the environment are a major factor in global biology. Early microbiologists won-dered at the fact that many pathogens and toxic substances end up in the soil and water, yet by and large, soil and water are relatively devoid of such germs and sub-stances. In some way they are detoxified and degraded. A remarkable early obser-vation by an English bacteriologist working in India, Ernest Hankin (1865–1939), was the fact that while it was the tradition to consigning the dead to the waters of the Ganges River, even during cholera epidemics, the Ganges did not seem to be the source of diseases at the level one might expect (19). This "Ganges paradox" had two different explanations: the traditional explanation was based on religious beliefs in the healing powers of the Ganges, the second explanation, investigated by Hankin, was that the Ganges water possessed some sort of antibacterial agent or substance that killed the cholera microbes constantly being dumped into it (20).

Human and animal waste processing became necessary as population densities increased with urbanization and by the sixteenth century collection of sewage in cesspits was introduced. The cesspits created primitive sewage treatment systems that would eventually evolve, via septic tanks, to the modern sewage treatment systems that utilize microbial action to process the sewage for eventual recycling into the environment. This recycling process is still monitored by a basic measure of microbial action, the uptake of oxygen by the microbial communities in the sewage as they oxidatively convert the complex organic matter in the sewage to more basic compounds. This measure of the "biochemical oxygen demand" or BOD is a simple and fairly reliable assay of the extent of microbial "sewage processing" (21).

The specific exploitation of the complex microbial processing of sewage has a long history, and none is more fascinating that of "Miloganite" (*Milwaukee organic nitrogen*) a commercial product introduced in 1927 to help the city of Milwaukee, Wisconsin dispose of its microbe-processed municipal sewage (22). The primary sewage treatment process involves a simple screening to remove large discarded objects that might interfere with the subsequent secondary processing. That processing involves simple settling to allow the solids to accumulate at the bottom of the settling tanks, and to allow the microbes to feed on the nutrients in the wastewater phase. After the BOD measurements indicate that the microbial degradation is complete, the treated wastewater is discarded into the local environment (Lake Michigan in the case of Milwaukee) and the sludge at the bottom of the tank is left to dry. The dry and pulverized sludge, rich in organic nitrogen compounds, is then ready for recycling as agricultural fertilizer (23). In the case of Milorganite, these solids are heated to destroy any potential pathogenic microbes, then packaged for sale.

While the global compartment we generally call soil is small compared to Earth's oceans, it, too, is of vital importance to life on the planet's surface, for example, for human beings. The soil, the surface of Earth to a depth of a few meters is not as microbe-rich as the deeper subsurface regions, nonetheless soil contains, on average, between 40 million and 2000 million bacteria in each gram of soil, the wide range reflecting quite different soil types and locations (24). Soil scientists have long appreciated the complexity of the microbial communities that live there as well as the roles these organisms play in agriculture and bioconversions of waste, both natural and man-made. Indeed, soil microbiology can be viewed as the grandparent of microbial diversity studies.

One obvious fact about soil biology is the process of conversion of myriad substances, both organic and inorganic into other substances of use by plants for their growth; various forms of "waste" such as dying animals and plants, manmade detritus such as paper and cardboard, as well as excreted residues of higher animals including humans. Less noticed is the conversion of inorganic substances containing iron, sulfur, and phosphorus into useful nutrients for the plant world. Even atmospheric nitrogen is trapped and changed in the soil. How does all this happen? The short answer is "microbes." These important chemical reactions that occur in soil were some of the earliest topics of interest to nineteenth century microbiologists when modern microbiology was born. These reactions were not only of scientific interest, but also of major agricultural and environmental interest. How do fertilizers work? What happens in sewage disposal? Do pathogens persist in the soil?

One vital process is that of the conversion in the soil of the nitrogen compounds in decaying material into different compounds suitable for uptake by growing plants. Nitrogen is an essential element of all proteins and all nucleic acids, and nitrogen is recognized as a main component of chemical fertilizers as well as of so-called "organic" fertilizers. In the nineteenth century soil scientists found that addition of various compounds with reduced nitrogen in the form of amino-groups, that is, $-NH_3$, resulted in the production of oxidized nitrogen in the form of nitrate, that is NO_3^{-2}, if the soil was incubated in the laboratory. Since soil that had been heated or treated with chloroform, a treatment that killed microbes, could not produce nitrate, they concluded that some form of life in the soil, most likely microbes, was carrying out this important reaction, one that they called nitrification.

Nitrification became the central focus of a brilliant, if somewhat eccentric, Ukrainian-born Russian, Sergei Winogradsky who we met earlier in Chapter 3. He studied first in St. Petersburg, but then finished his education in Strasbourg with Anton de Bary (1831–1888), one of the pioneers advocating for better understanding of the concept of symbiosis (25). This influence on Winogradsky was profound, and his work, first on nitrification, and later on broader aspects of microbial ecology paved the way for soil microbiology to become a leading area of microbiome interest. For some time Winogradsky puzzled over the microbial process of nitrification. His problem was one of elementary chemistry: oxidation of $-NH_3$ required two steps: one oxidation reaction to nitrite (NO_2^{-1}) and a second oxidation to nitrate (NO_3^{-2}), the form needed by plants. Winogradsky was following the dogma of his time to work with pure cultures in the laboratory, and the microbes he isolated would only carry out step one, producing nitrite. Finally, he realized what was needed was a second microbial species which could oxidize nitrite to nitrate, a microbe he eventually isolated. His insight, of course, was to realize that in soil, these two organisms were living together and cooperating to carry out the full conversion needed to produce the needed plant nutrient. Soil, he realized, harbored a complex, interacting microbial community. Although he retired early at age 50, after 15 years he came out of retirement in 1922 and joined the Pasteur Institute to found their laboratory of microbial ecology which he headed until his death in 1953. In his magnum opus on soil microbiology written near the end of his life, he emphasized that "...conditions of pure culture in an artificial environment is never comparable to that in a natural environment... [and that] ...one cannot challenge the notion that a microbe cultivated sheltered from any living competitors and luxuriously fed becomes a hot-house culture, and is induced to become in a short period of time a new race that could not be identified with its prototype without special study" (26). Winogradsky's pioneering views of the importance of this principle lives on in the form of the "Winogradsky column" the simple laboratory approximation of the rhizobiome in which soil, sludge, or other solid sample from nature is placed in a glass cylinder along with a source of carbon, sulfur, and calcium (cardboard, gypsum, and eggshells, for example) then sealed and exposed to sunlight for several months. Serial observations and eventual sampling show that different layers appear that represent variations in the microbial populations that evolve and become stable in the particular environments along the length of the column. The column equilibrates to have gradients of oxygen, hydrogen sulfide, and other components that result in the growth of diverse, interacting communities of microbes, many of which Winogradsky isolated and identified. In this way he investigated and developed many of our current ideas about microbial nutrient cycles in soil, diversity of interacting populations, and complexity of the rhizobiome, and this dramatic demonstration still lives on as a pedagogic tool coupled with metagenomic analysis to expand our understanding of such microbial communities (27).

Soil microbes, of course, play a key role in global food security. Over farming, poor fertilization practices, and climate changes have resulted in deteriorating soil quality, more desperate farming methods, and declining crop yields. Studies of some of these areas suggest that attention to the microbes in the soils might promote recovery of now barren land (28). Newly introduced microbes such as specific extremophiles may promote soil recovery where soil conditions have changed due to drought, elevated temperatures, salt accumulation, invasive pathogens, and even toxic oil spills. Some soil experts envision a new "precision agriculture" in which beneficial microbes could be prescribed for specific fields after an in-depth analysis of a field's rhizobiome (29). The adage, known to scientific gardeners for generations, that one "feeds the microbes not the plants" has become a reality more than ever.

HUMAN MICROBIOME

Salt water and coffee filters do not seem like the stuff of modern, high-tech medicine, but amazingly, they are the essential tools of a procedure called "fecal transplantation" that is having a significant resurgence after generations of relegation to jokes about "old time country docs." This practice in human medicine may have derived from the tradition in veterinary medicine where digestive disorders, especially in ruminants as well as horses, have been treated with digestive fluids from healthy animals in a procedure called "transfaunation" (30). Many nearly intractable intestinal ailments are being treated, often successfully, with filtered suspensions of feces obtained from healthy donors. Advances in metagenomics and understanding of the gut microbiome in health and disease now provides scientific support for these old-time practices, former called fecal enemas (referring to the route of administration of diluted fecal material from healthy sources) (31).

The microbiology of the human gut, a specialized microbial environment of the human body, has intrigued scientists and physicians almost from the dawn of the nineteenth century germ theories. At the Pasteur Institute at the start of the twentieth century, the already famous Ukrainian-born immunologist Elie Metchnikoff (1845–1916) turned his attention to the problem of aging. Why, he asked, did the peasants of Bulgaria seem to enjoy an unusually long lifespan? His answer: yoghurt. From this basic correlation he developed his theory of autointoxication, the production and absorption of toxins in the long intestine of humans, teeming with microbes fermenting the waste of our food into unhealthy substances. Yoghurt, with its high content of lactic acid, produced by "good" bacteria such as the organism now known as *Lactobacillus delbrueckii* subsp. *bulgaricus*, was an antidote to these toxins. While some considered advocating removal of sections of the colon to prevent this autointoxication, Metchnikoff advocated for a diet rich in lactic acid-producing microbes. *Lactobacillus acidophilus*, a more robust microbe, one of the bacteria responsible for souring milk, became known as the bacteria of long life. While Metchnikoff called this process "orthobiosis" later followers called these organisms "probiotics." For decades this autointoxication theory was dogma, and still holds sway in the obsession of many with the idea of "regular" bowel movements, leading to "the laxative habit," and the popularity of various nostrums to promote "gut health" (32).

The human intestine and its community of microbes has long been seen as a complex ecological niche, important for human health. The microbial processing of food and the production of essential nutrients by the microbes for absorption by the human host was recognized in the 1930s with the development of both microbial chemistry and nutritional biochemistry. An interesting and humbling story surrounds the discovery of the important vitamin, biotin. In 1916 it was found that feeding large amounts of raw egg white could be toxic to several mammalian species, including humans (33). This "egg-white" injury was eventually traced to the egg white-induced depletion of an essential substance produced only by plants and bacteria; the essential substance was identified as the vitamin now called biotin. In 1941 a protein in egg white (avidin) was identified that very tightly bound to biotin making it unavailable to the host organism. As it turns out, the bacteria in our gut is a main source of biotin, regardless of whether or not we eat enough of it.

With the introduction of potent antibiotics, surgery on the intestines became much safer since infections from contaminating fecal leakage could be suppressed by feeding high doses of non-absorbable antibiotics such as neomycin to effectively sterilize the gut prior to surgery. The side effect of this treatment, however, was disruption of the normal bacterial functions of the gut. One could simply wait out the gastrointestinal symptoms after surgery allowing the normal flora of the gut to re-establish itself over time from the patient's environment. Some physicians, however, routinely tried to repopulate the patient's gut with a mix of normal flora by prescribing enemas of diluted fecal material from presumably normal patients. The current exploration of fecal transplants, thus, really does represent old wine in new bottles (34).

The capacity of the gut microbiome to carry out a multitude of reactions, both producing and consuming substances in an organ, the gut, designed for transport of materials into and out of the body, suggests its importance in many regulatory and physiological processes. Signaling hunger and nausea, taking up nutrients, toxins, psychoactive substances, and immunomodulatory molecules, the chemistry and hence the microbes of the gut make it a prime target for understanding and manipulating microbiome populations. Metagenomic analyses of the gut microbiome have been correlated with diverse conditions in search of causal links. One particularly prominent area has come to be called the brain-gut axis (35). Correlations between disordered gut microbiome profiles and specific neurological and psychiatric diseases have been reported widely in the past decade or so. Sone of these correlations are based on plausible physiological knowledge, others are simply fanciful. Since there is strong empirical and theoretical support for certain neural pathways that affect gut function, for example, the control of gut motility by neural signals via the vagus nerve, from brain to gut musculature, it is not surprising that brain activity that slows or speeds up the transit times of bacterial populations through the gut might alter the microbial population profile in such cases. Patients with irritable bowel syndrome often have altered microbiome profiles that revert toward normal when their condition improves. Since this syndrome is thought to have a strong neural component, it is likely that this is also a concrete case of brain to gut influence, not the other way around (36). While there are isolated reports of, for example, improvement of mental symptoms when the microbiome is modified by antibiotic therapy or feeding specific probiotic species, these studies are almost uniformly not conducted to the rigorous standards to demonstrate any causal links. Placebo effects are particularly strong in many such reports as well (37). The studies to determine if there really are causal links between brain function and the gut microbiome are difficult to design and carry out, and the demonstration of the direction of interaction, brain on gut or gut on brain, will be essential as well. At present, the belief that "a healthy gut means a healthy brain" is mostly a nice advertising slogan, not a well-supported scientific truth.

There is one quite recent hopeful application of gut microbiome biology in medicine, however. The widespread use of broad-spectrum antibiotics, treatments that in many cases lead to reduction of the population size and diversity in the intestine as an untoward side effect, has opened the way for colonization of the gut by a noxious organism, *Clostridioides difficile* that causes troublesome diarrhea. While *C. difficile* is often treatable with more antibiotics (usually vancomycin), in the absence of competing organisms in the gut, the colonization often recurs. Research in the past several decades has shown that *C. difficile* is a poor competitor when confronted with a normal, healthy, diverse gut microbiome. Successful trials of fecal transplants in cases of recurrent *C. difficile* diarrhea are now establishing this old procedure as standard, some say "gold standard," treatment for recurrent *C. difficile* diarrhea (38).

The promise of personalized medicine through gut metagenomics has become a goal worth exploring now that metagenomic analyses are nearly affordable and becoming readily interpretable. Many gastrointestinal diseases, digestive symptoms, and nutritional problems are likely to become more understandable and treatable with help from metagenomics (39).

NOTES AND REFERENCES

1. **Whitman WB, Coleman DC, Wiebe WJ.** 1998. Prokaryotes: the unseen majority. *Proc Natl Acad Sci USA* **95**:6578–6583.
2. **Bar-On YM, Phillips R, Milo R.** 2018. The biomass distribution on Earth. *Proc Natl Acad Sci USA* **115**:6506–6511.
3. See Reference 2. The microbial world in this calculation includes the life forms usually claimed by the discipline of microbiology: bacteria, archaea, viruses, protists, and fungi. Bacteria and archaea comprised 83% of this total microbial biomass.
4. Bachman (1924-1999) received her Ph.D. from Stanford under the direction of van Niel in 1954.
5. **Spath S.** 2004. Van Niel's course in general microbiology. *ASM News-Am Soc Microbiol* **70**:359–363.
6. **Buitenhuis ET, Li WK, Lomas MW, Karl DM, Landry MR, Jacquet S.** 2012. Picoheterotroph (Bacteria and Archaea) Biomass Distribution in the Global Ocean. *Earth Syst Sci Data* **4**:101–106; see also **Cho BC, Azam F.** 1988. Major role of bacteria in biogeochemical fluxes in the ocean's interior. *Nature* **332**:441–443; and **Herndl GJ, Reinthaler T, Teira E, van Aken H, Veth C, Pernthaler A, Pernthaler J.** 2005. Contribution of archaea to total prokaryotic production in the deep Atlantic ocean. *Appl Environ Microbiol* **71**:2303–2309

7. **Webster NS, Reusch TBH.** 2017. Microbial contributions to the persistence of coral reefs. *ISME J* **11**:2167–2174.

8. **Margulis L.** 1991. Symbiogenesis and symbionticism, p 1–14. *In* **Margulis L, Fester R** (ed), *Symbiosis as a Source of Evolutionary Innovation: Speciation and Morphogenesis.* MIT Press, Cambridge, Mass.

9. **Knowlton N, Jackson JBC.** 2001. The ecology of coral reefs, p 39–422. *In* **Bertness MD, Gaines SD, Hay ME** (ed), *Marine Community Ecology.* Sinaur Associates, Sunderland, Mass.

10. See Reference 7, p 2168.

11. See Reference 7, p 2171.

12. **Mushegian AR.** 2020. Are there 10^{31} virus particles on Earth, or more, or fewer? *J Bacteriol* **202**:e00052–e20.

13. **Breitbart M, Bonnain C, Malki K, Sawaya NA.** 2018. Phage puppet masters of the marine microbial realm. *Nat Microbiol* **3**:754–766.

14. **Carlson MCG, Ribalet F, Maidanik I, Durham BP, Hulata Y, Ferrón S, Weissenbach J, Shamir N, Goldin S, Baran N, Cael BB, Karl DM, White AE, Armbrust EV, Lindell D.** 2022. Viruses affect picocyanobacterial abundance and biogeography in the North Pacific Ocean. *Nat Microbiol* **7**:570–580.

15. **Sims Woodhead G.** 1891. p 392. *In Bacteria and their Products.* Walter Scott, London, UK.

16. **Migula W,** 1890. *Die Artzahl der Bakterien bei der Beurtheilung des Trinkwassers:(Sep. Abdr. a. Centralblatt f. Bakteriologie u. Parasitenkunde Bd. 8) Habilitationsschriften der techn. Hochschule in Karlsruhe.* Fromann, Jena, Germany. Cited in Reference15, p 393.

17. **Brumfield KD, Hasan NA, Leddy MB, Cotruvo JA, Rashed SM, Colwell RR, Huq A.** 2020. A comparative analysis of drinking water employing metagenomics. *PLoS One* **15**:e0231210.

18. **Anantharaman K, Brown CT, Hug LA, Sharon I, Castelle CJ, Probst AJ, Thomas BC, Singh A, Wilkins MJ, Karaoz U, Brodie EL, Williams KH, Hubbard SS, Banfield JF.** 2016. Thousands of microbial genomes shed light on interconnected biogeochemical processes in an aquifer system. *Nat Commun* **7**:13219.

19. In 2017 the author witnessed several funeral rituals along the banks of the Ganges River in the religious city of Varanasi where remains of human funeral pyres were swept into the river adjacent to where children, businessmen, and water buffalo were swimming.

20. **Hankin EH.** 1896. L'action bactericide des eaux de la Jumna et du Gange sur le vibrion du cholera. *Ann Inst Pasteur (Paris)* **10**:511–523.

21. **Royal Commission on Sewage Disposal.** 1915. *The Final Report: Treating and Disposing of Sewage.* His Majesty's Stationery Office, London, UK.

22. **Schneider D.** 2012. Purification or profit: Milwaukee and the contradiction of sludge, p 170–192. *In* **Foote S, Mazzolini E** (ed), *Histories of the Dustheap: Waste, Material Cultures, Social Justice.* MIT Press Cambridge, Mass.

23. **Environmental Protection Agency, Office of Water (4204).** 1998. *How Wastewater Treatment Works... The Basics.* United States EPA 833-F-98-002.

24. See Reference 1, p 6579.

25. **Ackert LT Jr.** 2007. The "Cycle of Life" in ecology: Sergei Vinogradskii's soil microbiology, 1885-1940. *J Hist Biol* **40**:109–145. An alternate Romanization of his surname is Winogradsky.

26. **Winogradsky S.** 1949. *Microbiologie du Sol. Problèmes et Méthodes.* Masson, Paris, France. [Translation from **Dworkin M.** 2012. Sergei Winogradsky: a founder of modern microbiology and the first microbial ecologist. *FEMS Microbiol Rev* **36**:376].

27. **Rundell EA, Banta LM, Ward DV, Watts CD, Birren B, Esteban DJ.** 2014. 16S rRNA gene survey of microbial communities in Winogradsky columns. *PLoS One* **9**:e104134.

28. **de Vrieze J.** 2015. The littlest farmhands. *Science* **349**:680–683; see also **Ji M, Fan X, Cornell CR, Zhang Y, Yuan MM, Tian Z, Sun K, Gao R, Liu Y, Zhou J.** 2023. Tundra soil viruses mediate responses of microbial communities to climate warming. *mBbio* **14**:e0300922.

29. See Reference 28, **de Vrieze**, p 681.

30. The microbiology of ruminants (cattle, sheep, antelopes, deer, giraffes, and their relatives) has been extensively studied in veterinary medicine. Animals that feed mainly on plants have developed digestive systems to exploit microbial processing of the plant material into nutrients suitable for mammalian metabolism. To restore the microbiome of the sick ruminant, samples of the rumen contents from health animals are fed to the sick animal. Perhaps in a historical reference to Leeuwenhoek's "little animals" this procedure has come to be called "transfaunation" even though the microbes (mainly bacteria) being transferred are not considered animals (fauna). See **DePeters EJ, George LW.** 2014. Rumen transfaunation, *Immunol Lett* **162**:69–76.

31. de **Vrieze J.** 2013. Medical research. The promise of poop. *Science* **341**:954–957; see also **Kelly CR, Kahn S, Kashyap P, Laine L, Rubin D, Atreja A, Moore T, Wu G.** 2015. Update on fecal microbiota transplantation 2015: indications, methodologies, mechanisms, and outlook. *Gastroenterol* **149**:223–237; and **Waller KM, Leong RW, Paramsothy S.** 2022. An update on fecal microbiota transplantation for the treatment of gastrointestinal diseases. *J Gastroenterol Hepatol* **37**:246–255.

32. **Podolsky SH.** 2012. Metchnikoff and the microbiome. *Lancet* 380:1810–1811; see also **Mackowiak PA.** 2013. Recycling Metchnikoff: probiotics, the intestinal microbiome and the quest for long life. *Front Public Health* **1**:52

33. This is an outstanding reminder that the dictum that "natural is good" can be misleading. Raw egg white represents a truly "natural" and "unprocessed" foodstuff yet a food that can have detrimental health effects.

34. The author recalls his assignment as a medical student in the early 1960s to collect pails of soiled diapers from "normal" infants on the pediatric ward and flush them with saline to prepare fluids for such fecal enemas.

35. **Miller I.** 2018. The gut-brain axis: historical reflections. *Microb Ecol Health Dis* **29**:1542921.

36. **Whitehead WE, Palsson O, Jones KR.** 2002. Systematic review of the comorbidity of irritable bowel syndrome with other disorders: what are the causes and implications? *Gastroenterol* **122**:1140–1156.

37. **Martin CR, Osadchiy V, Kalani A, Mayer EA.** 2018. The brain-gut-microbiome axis. *Cell Mol Gastroenterol Hepatol* **6**:133–148.

38. **Yang L, Li W, Zhang X, Tian J, Ma X, Han L, Wei H, Meng W.** 2022. The evaluation of different types fecal bacteria products for the treatment of recurrent *Clostridium difficile* associated diarrhea: A systematic review and network meta-analysis. *Front Surg* **9**:927970.

39. **Walker AW, Duncan SH, Louis P, Flint HJ.** 2014. Phylogeny, culturing, and metagenomics of the human gut microbiota. *Trends Microbiol* **22**:267–274.

40. **Hiseni P, Rudi K, Wilson RC, Hegge FT, Snipen L.** 2021. HumGut: a comprehensive human gut prokaryotic genomes collection filtered by metagenome data. *Microbiome* **9**:165.

9 Emerging Diseases, Evolution, and the Microbiome

One theme that runs through this book is the principle of microbial diversity. Before we examine a few of the consequences of microbial diversity, it will be important to review just how diversity in the microbial world comes about. Bacteria were late to be recognized as genetic objects, only by mid-twentieth century being admitted to the world of organisms with genes, mutations, and (rarely) sexual reproduction. Their orphan status was the result of their lack of a microscopically visible nucleus with chromosomes, their lack of a sexual mating system, and the ambiguity surrounding the appropriate unit of the organism: the visible population, i.e., the culture or colony, or the microscopic individual cell (1). For the most part, bacteria and archaea reproduce without sex, that is, one cell by itself simply divides into two progeny cells that grow a bit to become identical copies of the original parent microbe. This means that there can be very large populations of cells that all derive from a single cell and are all identical (except for sporadic mutations, of which more is about to be said) to the original founder cell. This population is often said to be a clonal (Greek: *klonári*, twig) population since the population all stems from a single progenitor.

In the laboratory study of bacteria, one usually deals with massive populations, often grown from a single cell, a so-called pure culture. The behavior of such a culture is frequently taken to be a property of the individual organism itself. For example, as a population grows in a mass culture for a long time, the cells in the culture may change shape, or show other changes in appearance or behavior. When individual cells from this changed population are used to start new clonal populations, the resulting cultures sometimes show new properties. The early microbiologists

Magic Bullets, Miracle Drugs, and Microbiologists: A History of the Microbiome and Metagenomics,
First Edition. William C. Summers.
© 2024 American Society for Microbiology.

called this phenomenon "dissociation" suggesting that the long-term growth caused the population to dissociate into new forms of cultural behavior. The culture itself seemed to have hereditary properties. In the 1920s, the soon-to-be-famous author of *Microbe Hunters*, Paul de Kruif (1890–1971) (when he was still officially a lab bench microbiologist) showed that this change in the culture was simply the result of selection of a few more well-adapted bacteria that arose by random mutation and eventually dominated the culture under the new conditions. Darwin's evolution by natural selection in action (2).

From these early clarifications, heredity and its mechanisms in bacteria gradually took shape. Bacteria and eventually viruses were found to have genes that behaved in their most basic form just like genes in mice, corn, and fruit flies, the popular experimental organisms of the early geneticists. In the late 1930s it was observed that yeast cells, that usually reproduce like bacteria by simple fission into two progeny cells, could actually mate now and then (3). This microbial sex was important for two reasons: it suggested that sexual mating to bring two genetically different organisms to produce a new combination of genes in the progeny was probably an important evolutionary event, stimulating vastly more "genetic exploration" of possible evolutionarily more fit "higher" organisms, and second, that since the microbial genes operated by the same underlying mechanisms as do all other forms of life, their vastly superior populations meant that microbes comprise the largest reservoir of potential new genes on the planet.

In addition to gene mutation that was known in many well-studied organisms from fruit flies to humans, bacteria (and probably other microbes) have been found to acquire diversity by mechanisms that might be called gene invasion. Viruses infect cells by introducing viral genes into host cells where, for the most part, the viral genes direct the production of more virus, often doing damage to the host cell in the process. Sometimes, however, the viruses also carry in genes that take up permanent residence in the host cell and simply change the properties of the cell by these genetic additions. As we discussed earlier, another of the microbial surprises of the 1950s and 1960s was the discovery that several mechanisms of such "lateral" or "horizontal" gene transfer were both common and significant in bacteria (4). Many of these invading genes were maintained as small auxiliary sets of genes that were physically not connected to the main genes of the microbe on what came to be called the chromosome of the organism. In his review that synthesized much knowledge of these auxiliary genes, Joshua Lederberg referred to these DNA strands as plasmids, emphasizing their lack of connection with the main chromosome (5). Shortly afterward, as more examples of these extra, usually optional mini-chromosomes, the designation as "episomes" was also used, to emphasize their extra-chromosomal, optional nature (6). A central conclusion of modern microbial genetics located microbial life firmly in the realm of Darwinian evolution like all other forms of life. Now we can ask just what this means with respect to microbial diversity, ecology, disease emergence, and the new concept called the microbiome.

As the concepts of evolution and genetics were brought together in the mid-twentieth century in the so-called "modern synthesis," our ideas about evolution and what it entails became more nuanced. One important new principle emerged from the work of the British evolutionary biologist William D. Hamilton (1936–2000) who proposed a theory of "kin-selection" in 1964 to explain the paradox of altruism (7). He emphasized that reproductive fitness, the gold ring of organic evolution, was the result of diverse contingencies, including social and behavioral forces, that contributed to his concept of "total inclusive fitness." Many of the examples of microbial communities, holobionts, and symbiosis that we have discussed are now being examined with a view to their inclusive fitnesses. For example, to understand the emergence of "new" infectious diseases, it is not enough to consider just the isolated microbe. Instead, ecological selective pressures, gene invasions from episomes and viruses, and host and competitor evolution all need to be taken into account.

As concrete examples of how microbial diversity arises, is influenced by environmental contingencies, and impacts the populations, we will consider first, how the microbe of plague arose from a rather benign ancestor, how a latent microbe in fruit bats emerged as the result of ecological changes to produce the lethal outbreak of Nipah virus disease in Asia in recent times, and, finally, how genetic diversity in influenza impacts its global pandemic potential.

PLAGUE: MICROBIAL EVOLUTION AND A NEW DISEASE

Plague, frequently called "the Black Death" in reference to a pandemic that ravaged Europe and the Middle East in the mid-fourteenth century, is a serious disease of humans caused by infection with a bacterium, traditionally called the plague bacillus, and now known to microbiologists as *Yersinia pestis* (8). Historians and epidemiologists describe three great, nearly world-wide, pandemics of plague. The first pandemic has been called the Plague of Justinian in (dubious) honor of the Eastern Roman Emperor Justinian (ca. 482–565), frequently dated as starting in 541 in central Africa and spreading northward to engulf Justinian's Mediterranean world. The second, more familiar pandemic has been traced to its origin in Central Asia in 1347 when it then spread via Crimea across Europe and became known as the Black Death. This second outbreak was followed by nearly three centuries of smaller, localized epidemics recognized as descendants of this plague. The third pandemic started in 1894 in Southeast Asia, usually pinpointed to Yunnan province in China, first spreading to Hong Kong, then India, and subsequently becoming world-wide from there, carried by the international trade routes of the modern world.

The new science of microbiology was challenged to understand plague, its cause, its pathology, and its epidemiology. In Hong Kong, Alexandre Yersin (1863–1943), a scientist from the Pasteur Institute in Paris was able to identify a bacterium that proved to be the causative organism of plague; the bacterium now bears his name, *Yersinia pestis*. This major advance, however, did not explain the epidemiology of the plague pandemics. In India, another Pastorian, Paul-Louis Simond (1858–1947), demonstrated an important new concept in infectious disease science, the need in some cases for indirect transmission of the microbe from host to host. He found that the plague organism could be carried from infected animals (and presumably other humans) to human beings by the bite of a flea that had acquired the infection from feeding on the blood of another animal. In other words, the flea was the "vector" of the disease. This principle was crucial to advancing the understanding of epidemiology of several other contagious diseases as well, for example, malaria, yellow fever, and dengue, all of which use mosquitos as intermediate hosts for transmission of microbes to humans.

Insect vectors became an important factor in both the understanding as well as the control of infections and epidemics. If one could eliminate the vector one might prevent transmission to humans of microbes that reside in animal reservoirs. Direct killing the vectors or elimination of their habitats became a major public health goal. Complementary approaches to control or eliminate contact with animals that harbor microbes that might be harmful to humans is another key public health goal.

The Black Death occurred just at the time when printing, travel, and social organizations were developing in several regions of the world. For this reason, we have a rich historical record of the second pandemic that is largely missing for Justinian's pandemic. Plague pandemics soon became a "literary disease" (9). Even now, plague continues to be a favorite subject of historians, scientists, and horror story writers.

Plague has also captured the attention of the microbial geneticists with their new tool, metagenomics. Because DNA is a rather stable molecule, and when hidden away in protected areas of a body, enough DNA can be recovered, even from ancient skeletons, to allow genes of long-dead individuals to be sequenced. Teeth, which have tiny blood vessels still containing dried blood inside in the pulp of the tooth, have been used to recover ancient DNA for study of the genes of our ancestors (10). These DNA samples often include the genes of the microbes that were actively infecting the individual at the time of her or his death. We can, in a way, do a clinical microbiological study of ancient populations. If we can obtain samples with a known medical history, it is possible to reconstruct a historical account of their health status and exposure to microbes. If the DNA samples are sufficiently preserved, the genetic makeup of the microbes can often be determined.

Amazingly, just as we can determine the genetic diversity of the microbes in a sample of soil or sea water, we now can determine the precise genotype of an ancient pathogen.

It is known from historical records from many locations that during the Black Death, the many corpses that piled up were simply buried in mass graves, now called (rather insensitively) "plague pits." During the height of the epidemic, it is safe to assume that most of the bodies in these mass graves died of the Black Death. DNA sequencing of numerous samples from these plague pits confirm that the vast majority of these individuals were infected with modern plague microbes. The Black Death has now been confirmed by current medical science as a *Y. pestis* pandemic.

Interesting as this retrospective diagnosis may be, an even more amazing conclusion is emerging from the metagenomics of ancient plague DNA. We now have a pretty good outline of how plague microbes themselves evolved from a rather benign gut microbe into a bacterium of global lethality. All this started about 3000–4000 years ago in Central Asia and these evolutionary events show, in impressive detail, how modern metagenomics works to unravel the puzzles of microbial diversity and evolution.

The story of how the plague microbe sprang into being to become such a lethal scourge of humankind starts, like many scientific advances, with an obscure and unlikely observation. Two scientists in the histology laboratory at the College de France in Paris in 1883, Louis-Charles Malassez (1842–1909) and William Vignal (n.d.), were studying atypical cases of human tuberculosis by inoculating guinea pigs with patient materials. In this process they found that guinea pigs developed tuberculosis caused by a distinctly different microbe from that isolated as the cause of common human tuberculosis by the famous Robert Koch (11). This new bacillus, then known as the Malassez-Vignal bacillus was not at all like the Koch bacillus of human tuberculosis, but it was also found in many healthy animal species, only rarely associated with specific diseases. Subsequently this bacterium was called the bacillus of *pseudotuberculosis* (that is, false tuberculosis). A decade later, a bacterial relative of this organism was discovered in Hong Kong by young Dr. Yersin, of the Pasteur Institute, who was there to investigate an outbreak of plague. Yersin's microbe was called *pestis* (pest, a traditional term for plague). Both these organisms were assigned to the same genus based on their common appearance and cultural properties. This genus was later named *Yersinia*. While *Y. pestis* caused plague, *Y. pseudotuberculosis* caused guinea pig tuberculosis, occasional outbreaks of gastroenteritis in the West, and strange scarlet-fever like disease outbreaks in the Far East (12). For years, microbiologists wondered what this close relationship might portend. Then genomics clarified everything. It seems quite certain that the plague microbe is simply a robust, enhanced version of its not-too-distant ancestor, the tame pseudotuberculosis bacterium.

Metagenomic studies of DNA samples from many plague burial sites, from the three distinct global pandemic periods show that about 3000–4000 years ago the relatively harmless gut microbe, *Y. pseudotuberculosis* acquired new genes by lateral gene transfer, lost several genetic properties by gene mutations, and consequently gained crucial properties that turned it into a lethal pathogen that we know as *Y. pestis* (13). These specific properties include the ability to grow in the gut tissue of fleas, facilitating vector spread by an ubiquitous human parasite and several interacting mutations that subvert host immunity and allow the microbe to survive, spread, and multiply in the human blood stream, lymph nodes, and lungs. In the case of plague, we have detailed evidence tracing the entire process of "emergence" of a new disease that has occurred in (relatively) recent times. It is likely that similar scenarios have occurred in the past for other microbes, and will likely occur again in the future.

NIPAH VIRUS: ECOLOGY OF DISEASE

A particularly clear and interesting recent example of the role of ecological factors can be seen in the emergence of a new virus called Nipah virus, named for the village in Malaysia where the first infection was recognized in 1998. The outbreak was first seen on several pig farms and resulted in a neurological and respiratory disease in humans with 265 cases and 108 deaths, clearly a terrible virus. Some of the pigs also developed respiratory disease but with much lower mortality. Eleven cases with one death were soon observed among abattoir workers in nearby Singapore, a major consumer of Malaysian pork. To control the spread of the Nipah virus, it has been reported that a million pigs were killed; nearby Thailand suspended importation of Malaysian pigs. Since the identification of the virus, several other outbreaks in humans and pigs have been observed in Southeast Asia and India. The high lethality, apparently high transmissibility, and initial mysterious emergence has made Nipah a bioterrorism concern as well as the stuff of popular culture (14).

While the epidemiology and clinical features of Nipah are fascinating, from a microbiological view, the story of its emergence is even more interesting. Research has now shown that Nipah virus is normally present in several species of fruit bats and related flying foxes in the forests of Southeast Asia. These bats rarely have close contact with either domestic pigs or humans. The virus seems well-adapted to bats. In Hamilton's terms, its inclusive fitness provides sufficient virus replication in bats to continue to exist in the population but with minimal harm to the bat hosts. Nipah has adapted to bat biology. Pigs and humans are different; Nipah virus growing in these hosts causes significant pathology. How did Nipah find its way from bats and flying foxes into pigs and people? Much research (coupled with some plausible speculations) indicates the following scenario (15).

Starting in 1997 the conjunction of several unusual atmospheric events occurred in Southeast Asia that led to a reduction of the availability of flowering and fruiting trees in the native forests, especially in Malaysia. Haze from deforestation fires in Indonesia was exacerbated by a severe El Niño-driven drought. The foraging of the flying foxes and fruit bats that depend on these deep forest habitats resulted in these animals invading the cultivated lands adjacent to the forests. Some of these areas were near extensive date palm orchards where the palm sap was being collected and where free-range pig farming is practiced. The flying-foxes and bats foraged on the palm sap and at the same time contaminated pig watering vessels, thus introducing a novel microbe into the pig population (16). In the area around the Malaysian town of Ipoh an outbreak of pig respiratory illness was soon followed by human infections that were much more serious (Figure 9.1).

In contrast to the case of plague in which a new disease emerged because of genetic changes in a progenitor microbe, in the case of Nipah, a new disease emerged because of disruption of ecological conditions allowing the interaction of previously isolated species that prevented cross-species transmission.

The example of Nipah virus emergence also raises another interesting phenomenon: the apparently unusual role that bats and flying-foxes may play in virus emergence. These mammals that can fly like birds have long held a special place in the human imagination (17). Many species of bats are nocturnal which reinforced their association with mystery and danger. The sixteenth century discovery of New

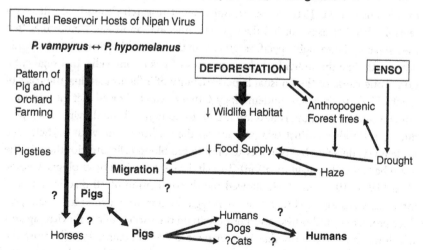

FIGURE 9.1 *Ecological interrelationships in the emergence of Nipahvirus in Malaysia. Drawn from Chua, 2003 (22), with permission.*

World blood-eating bats led to their being named as vampire bats to link them to ancient myths of vampirism. In more recent times, bats have been feared as frequent carriers of rabies virus. Fruit bats were identified as a reservoir for SARS coronavirus as well and more recently suspected as a reservoir for the COVID-19 virus (18). An apparently unusual feature of bat biology is their propensity to harbor multiple infections with microbes that are pathogenic to other mammalian species, yet show no symptoms themselves. Recent investigations seem to confirm that bats have evolved specific mechanisms to tolerate a heavy load of viral sequences and that the bats have a more complex relationship with viruses than previously thought (19).

INFLUENZA: GENETIC VARIATION AND EPIDEMICS

Influenza, a potentially severe respiratory infection, is the result of infection with virus belonging to the group of microbes known as orthomyxoviruses, an intimidating name that indicates that the virus is the usual or correct cause of mucus-producing disease. Influenza viruses that cause human "flu" epidemics also infect several other mammalian species, and, importantly, birds. In fact, many microbiologists consider "flu" to be a bird disease that also "spills over" to humans. The study of influenza provides a case study in the complexities of disease emergence, herd immunity, animal versus human epidemics, and the challenges of immunization programs.

Human experience with influenza presented a puzzle for a long time. Indeed, until the 1930s, the causative agent of flu was hotly disputed (20). Initially, attributed a bacterium called *Hemophilus influenzae*, in 1931 Richard Shope (1901–1966) identified a virus as the cause of swine influenza that was related to the cause of the 1918 epidemic. Subsequent research showed that this, and several related viruses, carried their genetic information in RNA molecules that were separated into eight specific segments in the core of the virus particle. While not exactly like the individual chromosomes of eukaryotic cells, this segmentation is one cause of the unusual hypervariability of influenza strains, the basis of much of flu's troublesome epidemiology. Other research showed that the way most humans and other animals develop immunity to fight off the flu virus is by producing antibodies against two proteins on the surface of the virus particle, one protein was detected by its ability to bind to red blood cells and clot them in the test tube (called hemagglutination). This clinical test, devised by virologist George Hirst (1909–1994) in 1942, allowed the discrimination of 18 distinct types of influenza virus proteins that cause hemagglutination to which animals can produce protective antibodies. Another protein on the surface of the flu virus against which we can make antibodies is an enzyme called neuraminidase (because the enzyme cleaves a sugar-like molecule called neuraminic acid to help release the virus particles from the infected cell). There are eleven known immunologically distinct types of neuraminidases known. Both these virus proteins are needed for a successful infection, the hemagglutinin (H) to attach and enter the host cell, and the neuraminidase (N) to break open the cell after virus replicates inside so the infection can spread. The good news is that our immune system can make potent antibodies against both proteins to combat flu infections. The bad news is that with 18 H and 11 N variants possible, the virus has a potential repertoire of 198 different variants with which to challenge us.

A virus like influenza, with eight genome segments, has some of the advantages of much more evolutionarily-advanced genetic systems such as eukaryotic cells: not just dependent on random mutations to generate diversity, these viruses can produce new genetic variants by reassortment of the segments. If two viruses with different versions of the H and different versions of the N genes happen to infect the same cell, the progeny viruses may have mixed segments during the process of genome replication and a new combination can emerge. A virus that might have type 1 H molecules and type 2 N molecules (H1N2, a predominantly pig strain) might meet up with another virus of the H3N8 type (infectious mostly in birds, dogs, and horses) and by reassortment, a new progeny strain of H3N2 type (Hong Kong 1960s human epidemic) might emerge. As it turns out, not all combinations are the same in terms of their biological properties. Some combinations efficiently infect only certain animal species to cause more or less severe symptoms. Some cause severe symptoms, but because their N protein function is not well-suited to that species, not much of the virus are released so its transmission rate is low. Other combinations may be just the opposite, highly transmissible, but with mild symptoms. Fortunately, of the 198 possible HN variants, only about 14 are currently known to cause significant human infections. Understanding the rather systematic variability of antigen reassortment in influenza had reduced a lot of the mystery and unpredictability of flu epidemics.

The reactions of our immune systems to flu viruses also influences the effects of these genetic variants. Flu epidemiology at first glance seems confusing and puzzling: why are there a lot more "scares" about bird and swine flu epidemics than there are actual human epidemics? Why is flu usually a serious disease only for the young and old? Why do we have to get an updated immunization yearly for flu but not for many other viral diseases?

Fortunately, mammalian immunity is usually rather long lasting; our antibody system is populated with so-called memory cells which have been selected by exposure to previous infections to have the genetic capacity to make useful protective antibodies for a long time. These memory cells are diverse because of our many diverse exposures in the past, but are few in number for any specific antibody. They are ready, however, when challenged again, to spring into action, rapidly multiply, and produce the specific virus-fighting antibodies (we observe this sometimes as swollen lymph nodes which are the site of such massive production of new antibody-producing cells). Such immunologic memory can last for years, sometimes for a lifetime. If we had been infected with the H3N2 strain in the early 1960s epidemic, commonly called the "Hong Kong" flu, many of us will still retain some memory cells and will be able to mount some, if not totally effective, resistance to this strain if we encounter it in 2023. Younger people, however, would be fertile soil, without prior exposure to this variant. This long-lived population immunity acts to dampen the population response to the emergence of variants that have occurred earlier in their lives. This is one of the explanations for the age distribution of susceptibility to flu. The young have no immunity to the reintroduction of past variants, and the old have waning immunity (21).

A logical consequence of this interaction between the natural reassortment driven variations in influenza virus and the immune system is that epidemics of such variants would be expected to occur in species and populations in relation to the life span of the population immunity. And this is indeed the case. While the average life span of a chicken under "normal" conditions is about five years (3–7 years), the vast majority of domestic chickens are not allowed to live out their lives to a natural death. Laying hens are said to live about 1.5–2 years while chickens for meat live 1.5–2 months. Thus, every year the world's chicken population becomes immunologically naïve and susceptible to every influenza virus that comes along. Some attempts to develop immunization strategies for chicken flocks have been undertaken, but so far with only mixed success. Epidemics of flu in chickens and swine, while economically devastating at times, are not directly indicative of a new wave of human disease. These animal epidemics are, of course, opportunities for new reassortments to take place, and more unpredictably, for more random mutations to occur. For the past several decades, microbiological understanding of genetic variability in flu viruses, coupled with genomic analyses of various animal reservoirs and known levels of human herd immunity, has allowed fairly accurate predictions of the viral variants likely to spread into human populations and cause serious pandemics (22).

CONCLUSION

The three examples of serious human infections described in this chapter show how new understanding of the genetics of microbial diversity can explain the origins of emerging infections (plague), the interplay of ecology, evolution, and culture (Nipahvirus), and the role of natural biological variations (influenza). All three of these cases display the result of the new molecular microbiological approaches to identifying, classifying, and analyzing microbes introduced in the decades following the genomic revolution of the 1970s. Indeed, most recently, we have seen how genome sequencing, genetically designed vaccines, and molecular epidemiology have been key to the rapid understanding and control of COVID-19.

Without these tools, products of the revolution in microbiology in the 1970s, it is hard to imagine how progress against this virus might have been made. It will come as no surprise, then, in the decades ahead many more microbes, diseases, and epidemics will become known to microbiologists as well as to the rest of us.

NOTES AND REFERENCES

1. **Summers WC**. 1991. From culture as organism to organism as cell: historical origins of bacterial genetics. *J Hist Biol* **24**:171–190.
2. de **Kruif PH**. 1921. Dissociation of microbic species II. Mutation in pure-line strains of the bacillus of rabbit septicemia. *Exp Biol Med (Maywood)* **19**:34–37.
3. **Winge Ø**. 1933. On haplophase and diplophase in some *Saccharomycetes*. *C R Trav Lab Carlsberg, Ser Physiol* **21**:77–108.
4. We usually think of genes being transmitted from parent to offspring, an example of "vertical transmission" between successive generations. When genes are transmitted between members of the same generation, say between sibling (avoiding the biology of organismal reproduction) it is called horizontal transmission" or "lateral gene transfer."
5. **Lederberg J**. 1952. Cell genetics and hereditary symbiosis. *Physiol Rev* **32**:403–430.
6. **Jacob F, Wollman E**. 1958. Episomes, a proposed term for added genetic elements. *CR l'Acad sci* **247**:154–156.
7. **Hamilton WD**. 1964. The genetical evolution of social behaviour. II. *J Theoret Biol* **7**:17–52. The paradox of altruism involves the continued existence of altruistic behaviors observed in many animal species when that behavior reduces reproductive efficiency. Naïve Darwinism would predict that such behaviors would be eliminated from populations over time. Hamilton suggested that altruistic behaviors, while selected against in the altruistic individual, could actually increase the reproductive fitness of genetically-related individuals in the social group. He emphasized the consideration of fitness factors that can interact, involve community members, and contribute to the "total inclusive fitness" of the species. For a recent view of total inclusive fitness see: **Dugatkin LA**. 2007. Inclusive fitness theory from Darwin to Hamilton. *Genetics* **176**:1375–1380.
8. *Yersinia pestis* is named for the discoverer of this organism as the cause of plague in Hong Kong in 1894, Alexandre Yersin (1863–1943) and a traditional term for plague, *la peste*. This organism was called *Pasteurella pestis*, but the genus *Yersinia* was first defined in 1944 and became official in 1980.
9. See for example: **Steel D**. 1981. Plague writing: from Boccaccio to Camus. *J European Stud* **11**:88–110; see also **Cooke J**. 2009. *Legacies of Plague in Literature, Theory and Film*. Palgrave Macmillan, New York, New York.
10. **Jones ED**. 2022. *Ancient DNA: The Making of a Celebrity Science*. Yale University Press, New Haven, Conn.
11. **Malassez L, Vignal W**. 1883. *Tuberculose zoogloeique*. *Arch Physiol Normale et Path* **2**:369–412; see also **Anonymous**. 1883. Recent researches on bacilli and tuberculosis. *BMJ* **18**:340–341.
12. **Amphlett A**. 2015. Far East Scarlet-Like Fever: a review of the epidemiology, symptomatology, and role of superantigenic toxin: *Yersinia pseudotuberculosis*-derived mitogen A. *Open Forum Infect Dis* **3**:ofv202.
13. **Demeure CE, Dussurget O, Mas Fiol G, Le Guern AS, Savin C, Pizarro-Cerdá J**. 2019. *Yersinia pestis* and plague: an updated view on evolution, virulence determinants, immune subversion, vaccination, and diagnostics. *Genes Immun* **20**:357–370.
14. The 2011 American movie *Contagion* (directed by Steven Soderburgh and written by Scott Z. Burns) is a medical thriller involving a global epidemic of a highly lethal and contagious virus emerging in Southeast Asia and resembling a hybrid of influenza and Nipah viruses.
15. **Chua KB, Chua BH, Wang CW**. 2002. Anthropogenic deforestation, El Niño and the emergence of Nipah virus in Malaysia. *Malaysian J Path* **24**:15–21; see also **Chua KB**. 2003. Nipah virus outbreak in Malaysia. *J Clin Virol* **26**:265–275.

16. **Sachan A, Singh R, Gupta B, Kulesh R.** 2023. Nipah virus and its zoonotic importance: a review. *J Entomol Zool Stud* **11**:208–213.

17. Bats and flying-foxes have adapted front limbs that have webs of skin that serve as wings much as the feathered fore-limbs of birds. They do not have an added set of wings in addition to the usual four appendages as imagined for angels and hippalectryons (e.g., Pegasus).

18. **Calisher CH, Childs JE, Field HE, Holmes KV, Schountz T.** 2006. Bats: important reservoir hosts of emerging viruses. *Clin Microbiol Rev* **19**:531–545. See also **Platto S, Zhou J, Wang Y, Wang H, Carafoli E.** 2021. Biodiversity loss and COVID-19 pandemic: the role of bats in the origin and the spreading of the disease. Biochem Biophy Res Comm **538**:2–13.

19. **Déjosez M, Marin A, Hughes GM, Morales AE, Godoy-Parejo C, Gray JL, Qin Y, Singh AA, Xu H, Juste J, Ibáñez C, White KM, Rosales R, Francoeur NJ, Sebra RP, Alcock D, Volkert TL, Puechmaille SJ, Pastusiak A, Frost SDW, Hiller M, Young RA, Teeling EC, García-Sastre A, Zwaka TP.** 2023. Bat pluripotent stem cells reveal unusual entanglement between host and viruses. *Cell* **186**:957–974.e28.

20. A good, readable account of these debates as well as a general history of the great 1918 epidemic is in **Barry JM.** 2005. *The Great Influenza: The Story of the Deadliest Pandemic in History.* Penguin, London UK.

21. Immunologists have proposed that for some flu variants young individuals with robust immunological responses can suffer severe symptoms related to the side-effects of their own immunobiology, a so-called "cytokine or immunological storm."

22. **Kilbourne ED.** 2004. Influenza pandemics: can we prepare for the unpredictable?" *Viral Immunol* **17**:350–357; see also **Agor JK, Özaltın OY.** 2018. Models for predicting the evolution of influenza to inform vaccine strain selection. *Hum Vaccin Immunother* **14**:678–683.

10 The Future of the Microbiome Concept

Although the concept of the microbiome has been around since the late 1970s, it has only come to wide-spread attention in the most recent decade since about 2010 (Figure 10.1). A central argument of this book has been that this concept owes its importance to a new understanding of microbial diversity revealed by new approaches in microbiology based on genetic analysis, evolutionary principles, and the technologies collectively known as metagenomics. No longer are microbes viewed as isolated, unwanted invaders, but rather as essential components of animal and plant physiology, sometimes even as essential symbionts. Large, diverse but stable communities of microbes are found in various physiological niches from deep sea corals to the human digestive track where they play crucial roles in health and disease. It seems that understanding specific microbiomes in all their complexities will only increase as a challenge in science and medicine.

In the early 2000s as scientists came to appreciate the interrelationships represented by the microbiome concept, they developed a view of the microbial world they called "One Health" to recognize the fact that many microbes were not limited to a single host or small group of hosts, but that they passed between many different species, each of which had specific interactions with a particular microbe. The One Health concept is not limited to infectious agents but was developed to meet "The need for a holistic, collaborative approach—One strategy to better understand and address the contemporary health issues created by the convergence of human, animal, and environmental domains is the concept of One Health" (1). While various authors have commented on the vague

Magic Bullets, Miracle Drugs, and Microbiologists: A History of the Microbiome and Metagenomics, First Edition. William C. Summers.
© 2024 American Society for Microbiology.

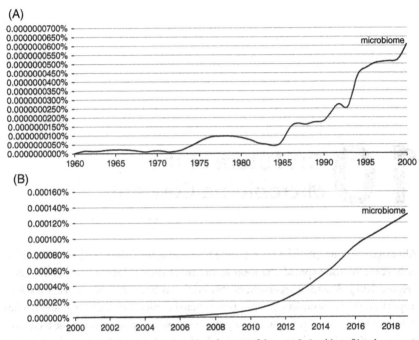

FIGURE 10.1 *N-grams (8) showing the prevalence, by year, of the term "microbiome" in a large English language data base; (A) 1969–2000; (B) 2000–2019; (a smoothing function has been applied to give the appearance of a continuous function).*

generalities of this aspirational concept, it gave support to the notion that microbes were not as compartmentalized by species, ecology, or pathologies as had been the general belief in earlier times. The new technologies and metagenomic analyses provided evidence to support the One Health concept for microbiology which coincided with the rise of the microbiome concept.

Scientists, always searching for novel ways to claim a discovery and often putting old wine in new bottles, have recast some of the core ideas of microbial diversity as components of Global Health or Planetary Health as a successor of the now "old" idea of One Health (2). Contrary to these linguistic gyrations, the microbiome concept seems more grounded in laboratory experimentation and secure ecological and evolutionary principles while dealing with many of the more interdisciplinary issues claimed by these other grand labels.

The community structures of specific microbiomes hold promise for future explanations of still poorly understood human, veterinary, and plant diseases. Recent metagenomic analyses in specific cases suggest that many physiological processes correlate with specific microbiome compositions. Some microbes seem to be beneficial, even protective while the presences of other microbes, of which many may not be pathogenic themselves, indicate potential risks to health.

We have already noted the importance of the gut microbiome, microbial communities long recognized as essential for normal physiology both in humans and other animals. Having the right mixtures of microbes for certain food processing such as wine and beer-making as well as fermentation processes involved in other such foodstuffs as leavened bread, cheese, yoghurt, pickles, kimchi, miso, and soy sauces has been known to generations of food technologists, cellar masters, cooks, and grandmothers. These and related processes, which sometimes go awry, will certainly benefit from the new knowledge derived from studies of the relevant microbiomes involved.

The impressive advances made possible by metagenomic technology in the understanding of the complex, interacting microbial communities we now call the microbiome, opens several new directions for medical progress. In the past, physicians suggested various approaches to altering specific microbiomes based only on the general notion that changes in certain microbial populations might be beneficial. Dermatologists advised using turmeric for eczema in the belief that some sort of immunological reactions were occurring, directed against certain skin microbes that would be suppressed by this treatment. Another traditional way to alter the skin microbiome thought to be involved in acne (via oily skin secretions) was to apply desiccating agents such as sulfur-containing lotions. The auto-intoxication theory of the early twentieth century was predicated on microbial production of "toxins" in the colon that were then absorbed into the blood stream. The treatment, short of surgical removal of most of the colon, was to promote frequent bowel movements, a practice that often led to dependence on laxatives to achieve "regularity" (3). These, and many other empirical or theoretical approaches to understand and exploit microbial diversity, are giving way to the use of microbiome studies as biomarkers for specific conditions of health and disease, as methods for disease surveillance, and as new ways to "hunt microbes." Let us consider just a few recent examples of this future promise.

It is clear that microbes provoke a variety of immune responses in animals and that often such immune responses include inflammatory reactions that produce local tissue changes, changes that can influence local microbiomes. An interesting and potentially important example of this interplay is the epidemiological observation relating HIV transmission and the circumcision status of one's male sex partner (4). Recent metagenomic studies of the microbiomes of circumcised and uncircumcised men suggest that circumcision alters the microbiome of the penis and as a consequence there is, on average, less inflammation and a lower level of HIV-carrying inflammatory cells. Since epidemiologic studies have found that circumcision reduces transmission rates by about half, it is being suggested as an important option for public health campaigns among non-circumcised men. Microbiome analyses may be expected to refine our understanding of this observation and its nuanced employment (5).

The recent COVID-19 epidemic has highlighted the utility of wastewater surveillance to monitor specific microbe loads in defined communities and populations. Routine expansion of such surveillance with metagenomic technologies is a likely future development in public health. Such monitoring will provide early warnings of known pathogens that find their way into the wastewater stream. Likewise, while insect-borne diseases such as Lyme disease are monitored in some high-risk areas, metagenomic monitoring will give a vastly expanded picture of such diseases if such monitoring becomes routine. Hospitals, too, may develop routine metagenomic surveillance for drug-resistance genes in environmental samples not dependent on culturing or patient sampling.

An interesting new way to hunt microbes, called "pathometagenomics" by its originators, may open the way to detect even more important microbial diversity. They hypothesize that disease emergence from animal populations may develop by new combinations of microbial genes, by new interactions between new hosts and altered microbes, and new combinations of selective pressures in new environments ... similar to the way we saw that plague originated several thousand years ago. In a first test of such a model for disease detection, they studied a natural mouse population as follows:

"Based on *in vivo* functional studies in inbred lab strains, it is hypothesized that the cost of prolonged bleeding times may be offset by an evolutionary trade-off involving susceptibility to a yet unknown pathogen(s). To identify candidate pathogens for which resistance could be mediated by B4galnt2 genotype [a mutation that affects blood vessels], we here employed a novel "pathometagenomic" approach in a wild mouse population, which combines bacterial 16S rRNA gene-based community profiling with histopathology of gut tissue. Through subsequent isolation, genome sequencing and controlled experiments in lab mice, we show that the presence of the blood vessel allele [gene] is associated with resistance to a newly identified subspecies of *Morganella morganii*, a clinically important opportunistic pathogen (6). In a similar approach to this hunt for unexpected diseases, recently a global surveillance strategy has been initiated to amass a repository of metagenomic screening data of blood samples collected from "mystery diseases." These data can then be correlated and searched for common features that will be applicable to both infectious conditions as well as other features that may point to underlying causes and conditions" (7).

The final and most widely anticipated application of human microbiome metagenomics is in the realm of what is now called "personalized medicine." Just as individual patients may soon be expected to carry a computer file with their entire genome sequence on a small card in one's wallet or uploaded on one's smart phone, periodic analyses of individual microbiomes from several parts of a patient's body may become routine. Diabetic patients now have nearly continuous monitoring of their blood sugar measurements, so it is not a stretch to think that

CartoonStock.com

FIGURE 10.2 *Journey to the center of Jules Microbiome. Source: Mark Heath/www.CartoonStock. com, used with permission.*

everyone might have periodic profiling of their microbiome spectra sampled from various crucial sites, for example, the intestines, oral cavity, genitalia, and sputum. As part of a routine health checkup, changes in these microbiomes may soon be part of the overall assessment of the state of health, as warning of incipient problems, and as part of a larger public health population surveillance (Figure 10.2).

IN CONCLUSION...

The understanding of the scope and diversity of our microbial world started with a few microscopic observations of interesting but randomly selected samples in the middle of the seventeenth century with the development of early microscopes. Such observations became more systematic, more extensive, and technologically improved in the two centuries that followed the discovery of these unseen forms of life. By the late-nineteenth century it became clear that many of these invisible organisms were involved in various important biological processes, some leading to diseases, others, as the basis for fermentation, putrefaction, and biodegradations of various sorts. The descriptions, classification, and biological relationships of the inhabitants of this microbial world were disputed, and murky. By the first half of the twentieth century however, improved techniques of chemical analysis, microscopy, and laboratory growth requirements led to general consensus on the identification, classification, and evolutionary relationships of most known microbes. They were finally "put in their place." A key aspect of this consensus was the belief that by mid-century most of the microbial world had been surveyed and the major outlines of its extent and diversity were then known. Various estimates suggested that scientists knew 95 percent of the inhabitants of the microbial world. This consensus held up until the decades of the 1960s and 1970s. At that time, as we have seen, several serious, new, and terrifying microbial diseases burst upon a rather complacent world. At the same time, a new technology, genetic analysis, including the beginning of direct attack on the DNA sequences of specific genes, would completely upset the hubris of the prior decades. In the decades that followed, right up to the present time, novel gene sequencing technologies have shown that microbiologists may know only about five percent of the organisms of the microbial world. The ability to determine gene sequences from large populations of microbes in parallel, a technique called metagenomics, has revealed the wide diversity of natural microbial populations and the importance of this diversity in ecology, evolution, and health. The community of microbes in a specific ecological niche has come to be called a microbiome. The microbiome concept has become useful to better understand the role of microbes in the biosphere and to provide a better understanding of our relation to these microscopic inhabitants of our world. And so as we look forward toward the future of where research into the microbiome and microbial diversity might take us, we do so with the weight of history and the wealth of experience from those who came before us, paving a new path forward without hubris.

NOTES AND REFERENCES

1. **King LJ, Anderson LR, Blackmore CG, Blackwell MJ, Lautner EA, Marcus LC, Meyer TE, Monath TP, Nave JE, Ohle J, Pappaioanou M, Sobota J, Stokes WS, Davis RM, Glasser JH, Mahr RK.** 2008. Executive summary of the AVMA One Health initiative task force report. *J Am Vet Med Assoc* **233**:259–261.

2. **Myers SS, Pivor JI, Saraiva AM.** 2021. The São Paulo declaration on planetary health. *Lancet* **398**:1299. See also **Horton R, Beaglehole R, Bonita R, Raeburn J, McKee M, Wall S.** 2014. From public to planetary health: a manifesto. *Lancet* **383**:847; and **Demaio AR, Rockström J.** 2015. Human and planetary health: towards a common language. *Lancet* **386**:e36–e37.

3. The famous microbiologist and one of the founders of modern immunology, Élie Metchnikoff actually advocated colonic resection as a way to promote longevity by preventing auto-intoxication. See **Kellogg JH.** 1917. Should the colon be sacrificed or may it be reformed? *J Am Med Assoc* **LXVIII**(26):1957–1959.

4. **Lin Y, Gao Y, Sun Y, Turner D, Zou H, Vermund SH, Qian HZ.** 2022. does voluntary medical male circumcision reduce HIV risk in men who have sex with men? *Curr HIV/AIDS Rep* **19**:522–525.

5. **Price LB, Liu CM, Johnson KE, Aziz M, Lau MK, Bowers J, Ravel J, Keim PS, Serwadda D, Wawer MJ, Gray RH.** 2010. The effects of circumcision on the penis microbiome. *PloS One* **5**:e8422; see also **Saxena D, Li Y, Yang L, Pei Z, Poles M, Abrams WR, Malamud D.** 2012. Human microbiome and HIV/AIDS. *Curr HIV/AIDS Rep* **9**:44–51.

6. **Vallier M, Suwandi A, Ehrhardt K, Belheouane M, Berry D, Čepić A, Galeev A, Johnsen JM, Grassl GA, Baines JF.** 2023. Pathometagenomics reveals susceptibility to intestinal infection by *Morganella* to be mediated by the blood group-related B4galnt2 gene in wild mice. *Gut Microbes* **15**:2164448.

7. **Aizenman N.** 2023. Scientists race to detect new pathogens before they can spark another pandemic. *NPR* 2 February. https://www.nprillinois.org/2023-02-02/scientists-race-to-detect-new-pathogens-before-they-can-spark-another-pandemic

8. **Google Books Ngrams.** https://registry.opendata.aws/google-ngrams. Accessed 6 November 2022.

Index

Magic Bullets, Miracle Drugs, and Microbiologists: A History of the Microbiome and Metagenomics,
First Edition. William C. Summers.
© 2024 American Society for Microbiology.